Five Questions in Biodynamic Agriculture

'I looked and this is what I saw'

Glen Atkinson PhD

© Garuda Consultants Ltd
16 April 2025 - All rights
Can be used for educational purposes

ISBN 978-0-473-74518-9
pdf   978-0-473-74519-6

Contents

Equisetum and the Kidneys	4
Source material	62
The Three Worlds in Biodynamics	80
Planets in the 3 Worlds	103
References	104
The Spirit in Biodynamics	105
References	140
Cell Salts in Biodynamics	141
Hauschka and Julius	181
Glenological Chemistry image	197
Conclusion	198

Equisetum and Fungus
v10 April 2025

A study of what Dr Steiner's medical lectures
say about Equisetum and Kidney function
in regard to the Agriculture Lectures

by Glen Atkinson

Contents

Introduction	5
What do we have	7
Equisetum and the Kidneys	16
The Field of Play	18
What are the Kidneys for the Human	29
What disturbs the Kidneys	30
Quartz and Silicic Acid	32
Why is Equisetum appropriate	33
How does this translate to BD	38
Hot and Dry Fungus	41
Powdery Mildew > Seasons	46
Clay — AlSi	47
500 and 501	48
Equisetum and 501	50
Equisetum and Bidor	54
Potency Choice	55
Equisetum Planetary Ruler	55
Epilogue	61
Appendix 1 - Original Texts	62

Introduction

The question of fungal problems within Biodynamic management remains a unresolved issue in most enterprises. The grape industry still relies on Sulphur sprays, the apple industry on Lime Sulphur, while Biodynamic stone fruit are non existent in many parts of the world.

The universal 'hope' for BD fungal control is Equisetum, however very mixed results are reported. Some people say it works well and others say it has no effect. It is used for ALL fungus, rots as well as powdery mildew and rusts, without any discrimination of the very different causes of these problems.

There is very little Biodynamic literature, of depth, on Equisetum, and I have seen none that has taken up Dr Steiner's indication to look at how Equisetum works on the human kidney for the answer. So I went looking. In RS's medical lectures there were several comments on Equisetum and the kidneys, indeed one gets the impression it was one of his favorite examples of illness. Interestingly though, his comments were always different from one time to the next, always coming from a different direction. The medical community have put some effort into understanding his comments, and they have a fairly coherent understanding of Equisetums role with the kidneys. However they have not taken the next step of applying it to horticultural disease.

There are many strings to pull together in this story, and rather than have you wade through the diverse quotes by RS, I will tell the story as I understand it, and place most of the quotes as an appendix, so you can read them afterwards. Firstly, though I will present the existing references from the Agriculture Course to establish our starting point.

Note: All RS quotes, in my text, are in italics. I minimally edit his quotes to make them clearer, and these additions are in a normal typeface.

What do we have?

While talking about the preparations in the Agriculture lectures, RS made the following comment in Lecture 5

"Healing is not based on the microscopic changes in tissues and cells, but on a knowledge of the larger connections; this must also be our attitude to the plant nature. And since plant nature is in this respect simpler than that of the animal or man, so its healing is a more general process and when sick it can be healed with a kind of "cure-all" remedy. If this were not so, we should often be in a fix with regard to plants, as we are with animals, though not with human beings. For a man can tell us where he feels pain. Animals and plants cannot; and it is fortunate that. here the curative process is almost the same for all plants. A large number of plant diseases (although not all of them) can really be arrested as soon as they are noticed by a rational management of our manuring - namely in the following way:

We must then add calcium to the soil by means of the manure. But it will be of no use if the calcium is not applied in a living condition. If it is to have a healing effect it must remain within the realm of the living. Ordinary lime or the like is of no use here. Now we have a plant which is very rich in calcium - seventy-seven per cent, of its substances is calcium albeit in very fine distribution. This is the oak and more especially its bark. In the bark we have something which is at an inter-mediate stage between plant and living earth. You will remember what I said to you about the kinship between bark and live earth. For calcium as required in this connection the calcium structure in the bark of the oak is almost ideal. Calcium in a living state (not dead, though even then it has an effect) has the property which I have already described to you: it restores order where the etheric body is working too strongly so that the astral element is prevented from reaching the organic substances. Calcium kills (damps down) the forces of the etheric body and so sets free those of the astral body. This is characteristic of all limestone.

But if it is necessary for an over-powerful etheric element to be damped down and contracted in a regular way - not suddenly nor jerkily so that shocks are produced - but in a steady and orderly fashion, we should use calcium in the particular form in which it is to be found in the bark of the oak tree."

In lecture 6 he says

"There remains for us one more subject to consider: the so-called plant diseases. Actually this is not the right word to use. The abnormal processes in plants to which it refers are not "diseases" in the same sense as are those illnesses which afflict animals. When we come later on to discuss the animal kingdom, we shall see this difference more clearly. Above all, there are not processes such as take place in a sick human being. For actual disease is not possible without the presence of an astral body. In man and animals, the astral body is connected with the physical body through the etheric body and a certain connection is the normal state. Sometimes, however, the connection between the astral body and the physical body (or one of the physical organs) is closer than would normally be the case; so if the etheric body does not form a proper "cushion" between them, the astral intrudes itself too strongly into the physical body. It is from this that most diseases arise.

Now the plant does not actually possess an astral body of its own. It does not therefore suffer from the specific forms of disease that occur in men and animals. This is the first point. The next point is to ascertain what actually causes the plant to be diseased. Now, from everything I have said on this subject, you will have gathered that the soil immediately surrounding a plant has a definite life of its own. These life forces are there and with them all kinds of forces of growth and tender forces of propagation not strong enough to produce the plant form itself, but still waiting with a certain intensity; and in addition all the forces working in the soil under the influence of the Moon and mediated through water. Thus certain important connections emerge, in the first place you have the earth, the earth saturated with water. Then you have the moon. The moon beams, as they stream into the earth, awaken it to a certain degree of life, they arouse "waves" and weavings in the earth's etheric element. The moon can do this more easily when the earth is permeated with water, less easily when the earth is dry. Thus the water acts only as a mediator. What has to be quickened is the Earth itself, the solid mineral element. Water, too, is something mineral. There is no sharp boundary, of course. In any case, we must have lunar influences at work in the earth. Now these lunar influences can become too strong. Indeed this may happen in a very simple manner. Consider what happens, when a very wet spring follows upon a very wet winter. The lunar force enters too strongly into the earth, which thus becomes too much alive. I will indicate this by red dot's. (See Diagram No. 11). Thus if the red dots were not here, i.e. if the earth were not too strongly vitalised by the moon, the plants growing upon it would follow the 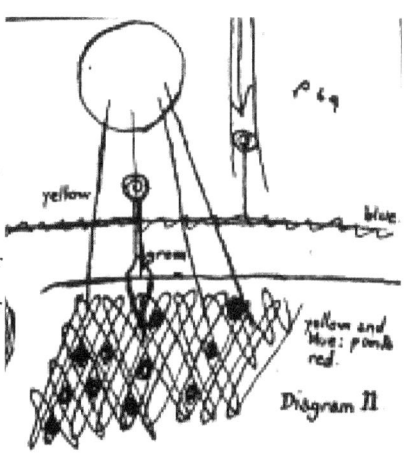 *normal development from seed to 'fruit; there would be just the right amount of lunar force distributed in the earth to work upwards and produce the requisite fruit seed. But let us suppose that the lunar influence is too strong - that the earth is too powerfully vitalised - then the forces working upwards become too strong, and what should happen in the seed formation occurs earlier. Through their very intensity the forces do not proceed far enough to reach the higher parts of the plant, but become active earlier and at a lower level. The lunar influence has the result that there is not sufficient strength for seed formation. The seed receives a certain portion of the decaying life, and this decaying life forms another level above the soil level. This new level is not soil, but the same influences are at work there. The result is that the seed of the plant, the upper part of the plant becomes a kind of soil for other organisms; parasites and fungaloid formations appear in it. It is in this way that blights and similar ills make their appearance in the plant. It is through a too strong working of the moon that forces working upward from the earth are prevented from reaching their proper height. The powers of fertilisation and fructification depend entirely upon a normal amount of lunar influence. It is a curious fact that abnormal developments should be caused not by a weakening but by an increase of lunar forces. Speculation might well lead to the opposite conclusion. Looking at it in the right way shows that the matter is as I have presented it. What, then, have we to do?* **We have to relieve the earth of the excess of lunar forces in it. It is possible to relieve the earth in this way. We shall have to discover something which will rob the water of its power as a mediator and restore to the earth more of its earthiness, so that it does not take up an excess of lunar forces from the water.** *This is done by making fairly concentrated brew (or tea) of equisetum arvense (horse-tail), diluting it and using it as a liquid manure on the fields for the purpose of fighting blight and similar plant diseases. Here again only small quantities are required; a homeopathic dose is generally sufficient. As you will have realised, this is precisely where one sees how one department of life affects another. If, without indulging in undue speculation, we realise the noteworthy effects produced by equisetum arvense upon the human organism by* **affecting the function of the kidneys**, *we shall have, as it were, a standard by which to estimate what this plant can achieve when it has been transformed into liquid manure, and we shall realise how extensive its effects may be when even quite a small quantity is sprinkled about without the help of any special instrument. We shall realise that equisetum is a first-rate remedy. Not literally a remedy, since plants cannot really be ill. It is not so much a healing process as* **a process exactly opposite to that described above.**

(lec 6 Ag course)

It is fair to say, both these passages are talking of fungal problems that arise from too strong an etheric activity coming from the

Earth and moving upwards, which suggests predominately fungal rot disease.

A few questions arise from his comments on fungal attack.
A) How can excess vitality lead to fungal attack
B) What is the effect of equisetum upon the kidney
C) How is equisetum the opposite of the excess vitality process RS described.

We can answer (A) by acknowledging, 'excess vitality', means excessive life and growth forces which come about from an overly active Etheric activity, carried in the excessive water, coming initially from the soil, but supported by a moist atmosphere.

Before we look at the other two questions, we need a broader overview, as Dr Steiner's medical view of life's organisation is more 'sophisticated' than is commonly used within Biodynamics.

RS made another comment about plant disease in lecture 1 of his 1921 medical lectures, that needs considering. There he said — Fungal attacks, can occur from the opposite direction. Human *'phenomena such as diphtheria are especially able to teach us about certain subtleties in the* plant *organism. Such diseases should be studied more precisely, if only for the sake of discovering remedies.*

In another context I have indicated that the child's acquisition of speech is accompanied by various organic processes. While he is learning to speak, and therefore while something special is taking place in his breathing organism, something also occurs polarically in his circulatory organism, which also receives into itself the metabolic processes. I also pointed out, how what at puberty appears in a reciprocal relationship of the human being to the outer world, takes place inwardly in learning to speak. Thus this push of the astral body, which at puberty takes place from within the human being outward, takes place from below upward in the capacity for acquiring speech. (from the metabolic towards the nerve sense) *So here we have an astralizing process, and we will be able to see clearly that an interaction occurs where the respiratory and circulatory systems meet (see drawing). The astralizing process working from below upward (yellow) encounters the developing organs of speech working from above downwards (red). In this encounter the organs of speech become stronger in their capacity for speech. It is what is taking place simultaneously below in the metabolism, that especially interests us here: this tends to work upward. The whole process is one from below upward (yellow arrows). Now, if the astrality presses upward too strongly while the child is learning to speak, we have a predisposition to diphtheric conditions. It is certainly important to pay proper attention to this.*

Let us now consider the outer earthly process, we see with plants, that has a certain selective affinity for the process I have just described. Let this be the surface of the earth . In a plant that behaves appropriately in relation to the cosmos, the earth plays a part in the nerve sense formation of its roots. With growth the influence of the earth diminishes and the extra-terrestrial influence becomes stronger and stronger, unfolding especially in the blossoms (see drawing, red). What develops here is a kind of external astralizing of the blossom, which then leads to the formation of fruit. If this process, which ought to occur in the normal course of the world processes, takes place below , it can only insert itself into the water, and we have what I have just called "dysentery of the earth."

But we can also have another

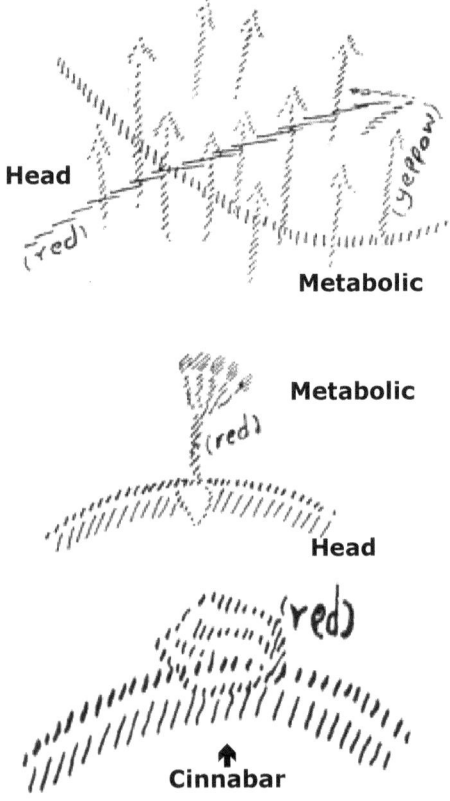

situation: What takes place when a plant develops properly — the blossom unfolding always a little above the earth's surface — can develop right on the earth's surface (see drawing below, red). Then fungi arise; this is the basis for fungus formation.

And now you will begin to guess that, if fungi arise from such a special astralizing process, the same process must take place from below upward when, as in human diphtheria, this remarkable astralization occurs in the human head. This is actually the case. Hence you find in diphtheria the tendency to fungoid formations. It is most important to consider this tendency to fungaloid formations in diphtheria, and it will also show you that a truly occult process is taking place there. Everything external is really only a sign that irregular astral currents, from the metabolism towards *the head,* are prevailing within the human being.

But when, as here, the processes work so deeply into the organism, much more will naturally be achieved by trying to find the specific remedy with which to oppose the particular process at work. One should try intermediate potencies of cinnabar. In cinnabar we will find effects that counteract all the phenomena I have mentioned. Cinnabar expresses this even in its outer appearance. If we acquire a sound understanding of such things we will recognize that cinnabar through its vermilion color is something that in a certain way brings to expression this activity opposed to the fungoid process. That which is approaching the colorless can become fungoid. While too strong an astralization of the earth's surface plays a part in the formation of fungi, in cinnabar there is a counter-reaction to this astralization and thus this reddening. Wherever a reddening appears in natural processes, we find a powerful counter-effect to the astralization process. You could express this in a moral formula: "The rose in blushing works against astralization." These domains of pathological-therapeutic study are really interconnected in a certain way. They guide us into this peculiar relationship of the ego and astral body to the other organs, to their laying hold of organs, to their emancipation from organs, or to manifestations of the excessive working of the astral from below upward in the human, which for the plant is from above downwards. (4)

Here we see fungal problems being described from a different direction. Not as a result of too strong an Etheric activity, but coming from too strong an astral activity. The solution is also different. Not equisetum by Cinnabar — Mercury Sulphate. This story reminds me of another from lecture 6 in the Agriculture course.

Now I am going to tread on very thin ice and take an example very near home. I am going to talk about the nematode of the beetroot. The outer signs of this disease are a swelling of root fibers and limpness of the leaves in the morning. Now we must clearly realise the following facts: The leaves, the middle part of the plant which undergo these changes, absorb cosmic influences that come from the surrounding air, whereas the roots absorb the forces which have entered into the earth and are reflected upwards into the plant. What, then, takes place when the nematode occurs? It is this: The process of absorption which should actually reside in the region of the leaves has been pressed downwards and embraces the roots.

Thus if this (Diagram No. 10) represents the earth level, and this the plant, then in the plant infested with the nematode the forces which should be active above the horizontal line are actually at work below it. What happens is that certain cosmic forces slide down to a deeper level; hence the change in the external appearance of the plant. But this also makes it possible for the parasite to obtain under the soil (which is its proper habitat) those cosmic forces which it must have to sustain It (the nematode

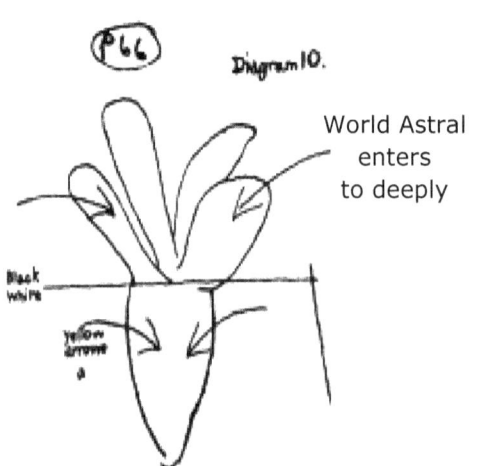

is a wire-like worm). Otherwise it would be forced to seek for these forces in the region of the leaves; this, however, it cannot do as the soil is its proper environment. Some, indeed all, living beings can only live within certain limits of existence. Just try to live in an atmosphere 70 degrees above or 70 degrees below zero and you will see what will happen. You are constituted to live in a certain temperature, neither above nor below it. The nematode is in the same position. It cannot live without earth and without the presence of certain cosmic forces brought down into it. Without these two conditions it would die out.

Every living being is subject to quite definite conditions. And for the particular beings with which we are dealing, it is important that cosmic forces should enter the earth, forces which would ordinarily display themselves only in the atmosphere around the earth. Actually the workings of these forces have a four-year rhythm. Now in the case of the nematode, we have something very abnormal. If one enquires into these forces, one finds that they are the same as those at work on the cockchafer grubs; and as those, too, which bestow on the earth the faculty of bringing the seed potato to development. Cockchafer grubs as well as seed potatoes are bred by the same forces, and these forces recur every four years This four yearly cycle is what must be taken into account not with regard to the nematode but with regard to the steps we take to combat it.

In both these cases we have an overly active World Astral activity causing two different manifestations. On one hand fungus, but what type, and on the other pest attacks, again what type and where? In the case of the pests RS did not provide the solution. In the Agriculture Course he went on to talk of peppering. However this does not provide a solution to this problem. Peppering does not alter the environmental energetic activities. It cuts off a specific reproductive stream to a specific part of a specific species. Experience shows , you might rid yourself of one pest via peppering, but another of the same natural niche will replace it. His suggestion for altering the environment to fend off pests is achieved by other means, such as specific BD preps and specific chemical elements. Nevertheless he has given us this very important image about the energetic background to some pest attacks, and that they have fungal relations.

The first three passages on fungus, all present different images and different solutions, for this subject.

Equisetum and the Kidneys

So lets start with the big hurdle - ***"we realise the noteworthy effects produced by equisetum arvense upon the human organism by affecting the function of the kidneys......... It is not so much a healing process as a process exactly opposite to that described above."***

What is RS view of the role of the kidney, and how does Equisetum do its job. Then how is all this opposite to the processes of excess vitality rising up from the soil.

The RS medical community has much to offer us. As a precise' of this we have this from the present co-leader of the Anthroposophical Medical Section, Matthias Girke. On page 873 of his book, 'Internal Medicine' (note: Anabolic = expansive, upbuilding, Catabolic = contractive, breaking down)

"Rudolf Steiner mentions Common horsetail in different instances as a medical plant for kidney diseases. Its purpose is to support the anabolic forces of the kidney organisation. Equisetum relieves the astral body of the disease process, allowing it to focus on the healthy kidney function.

Equisetum is characterized in terms of its substance by the polarity of sulfur and silicon.
Silicon is absorbed by means of the strong flow of liquid that is extracted from the locations on fields and near trails. This is deposited in epidermal cells at the top of the main shoot and the silicon content increases with the age of the plant. In connection with this silicon process, lens-like bulges can develop that direct the light `onto the

chlorophyll that is arranged in rows: In comparison to this, sulfur appears to occur `in a mineral- like dissolved form'. Equisetum thus encompasses a substance polarity that is related to the neurosensory system via the silicon and to the system of metabolism and limbs via the sulfur. These qualities can also be discovered in the threefold organization of the nephron in the form of the glomerulus, which is responsible for excretion processes, and the tubular system of the kidneys, which carries out the anabolic resorption processes. **Equisetum thus directs the light-like qualities of the neurosensory pole into the anabolic etheric area.** *Similar to the way in which sunlight is absorbed via the sensory organization in children and directed into the formation of the body and into their osseous metabolism, Equisetum* **appears to guide the Ego and astral entities into the anabolic life processes**. *This therapeutic gesture becomes immediately visible due to the formative processes of this medicinal plant that combine with the silicon and are carried into its life processes that are carried by the fluid substances. Equisetum is used specially for sclerotizing kidney diseases.*

In these cases, the astral organization orients itself too much towards the neuro-sensory pole and needs to be `repatriated: This can occur via the relieving therapeutic principle since the special nature of this plant, which exhibits a solidification in its silicic- mineral structure, allows it to take over the concordant disease process in the kidney. By means of the sulfur process, it reconnects the astral and Ego entities with the organism again. Rudolf Steiner mentions both this relieving principle and the efficacy of Equisetum in supporting the anabolic processes."

This passage provides the insight that Equisetum *"reconnects the Astral and Ego entities with the organism again" "Equisetum appears to guide the Ego and astral entities into the anabolic life processes", "Equisetum thus directs the light-like qualities of the neuro-sensory pole into the anabolic etheric area."*

This passage broadens our perspective somewhat, and when investigated further challenges the simplistic BD threefold view of the world to go a couple of steps further. The 'modern BD' image of Astral Above and Etheric Below, would interpret this passage to mean— Equisetum will draw the Astral from above more strongly, into the overly exuberant Etheric coming from below.

It may indeed be this simple, however the medical view presents a more complicated story than this. A closer reading of this passage shows all the energetic activities described are taking place within the metabolic region, with the astral and etheric balancing being within the kidney itself, and not primarily between the Head and the Belly. Also the end result of Equisetum is to stimulate the expansive Etheric activity, and it is the strong Etheric in the metabolism that would push off any nerve sense intrusion from the head. So would not strengthening the Etheric make the fungal problem worse? So questions arise.

The primary question also remains - What disturbed the metabolic astral's natural function, within the kidneys, in the first place?

To grasp the answer to this question, we need to gain an image of RS's overall worldview, and the specific context within which kidney disease occurs. To do so we need to expand 'modern Biodynamics' frame of reference somewhat.

The Field of Play

RS descriptions are based upon his view of how creation occurs, and how it then organises within living forms. So first we need to do a quick run through of 'the field of play', upon which we can see what he says.

The basis of RS thesis is 'As Above, So Below'. Whatever manifests as Life on Earth is an expression of the realities that exist within the cosmos above us. So to see life in its most honest sense we have to honour the truths of Astronomy.

The essential features we can identify is that we exist within an enormous Galaxy, within which there is a minute Solar

System, within which there is one of nine planets – the Earth, upon which we exist. But most remarkably, surrounding the Earth is an atmosphere, whose unique 18% oxygen density, has been formed

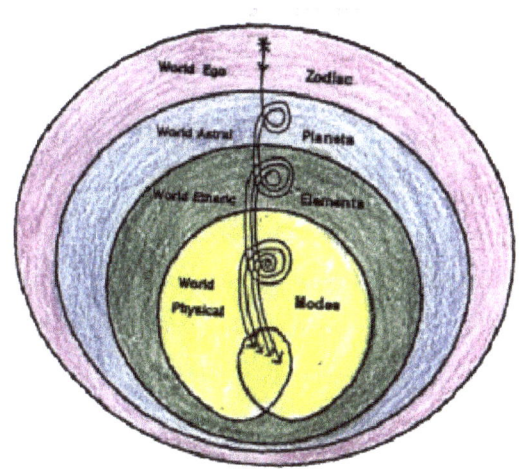

The Cosmic Ether

from the living process of the Earth itself. Hence we have four great astronomical spheres. This view only describes the substances we see. All of these substances organise within these organisations because of the electronic nature of these moving bodies. Electronics organise according to laws of polarity, which cause attraction and repulsion amongst everything created within these spheres of activity.

The Earth, while having its own energetic activities, also acts as a receiving organism for ALL of the electronic forces coming from above. RS called this complex, of galactic, solar system and atmospheric activities, **the Cosmic Ether**.

Astronomy tells us many things about the reality of these 'above' activities. The Galaxy is made up of trillions of stars. Our Sun is a star, and like our Sun all these other stars are beaming immense forces in every direction, for billions of years. The Solar System exists around the activity of this star. The Sun draws in forces and substance into itself before beaming out both of these things, out into the 'body of the solar system', along the horizontal plane. Due to the electronic organisation of the solar system, most of the substance in the solar system is vibrated to the middle of the horizontal plane of the sphere, where it begins to accumulate. Due to the spinning movement of this substance, and the Sun's Electronic field these accumulations of cosmic dust collect into what we now identify as planets. They are really glorified compost heaps. Nevertheless, we need to identify that there are nine of these planets, all in their designated places within the electronic onion that is the Solar system.

The Earth itself is spinning and has created its own electronic field, which extends out to the edge of the Ionosphere. Within this sphere there are bands of 'onion' like electro-magnetic organisation we identify as the Stratosphere, Troposphere etc.

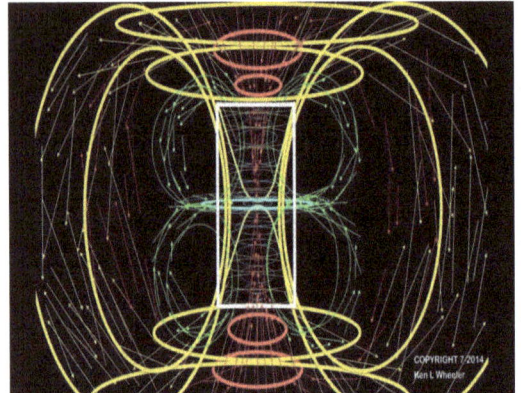

Given everything is electronic and actively energetic, all of these spheres interact with each other. As the stars are the most energetically active, their forces become dominant. They ray out in every direction, so when we look out from the Earth, we receive forces coming from stars, which reach us after journeying through and picking up the activities of the solar system onion. These combined forces then move through and pick up the forces from the atmosphere's energetic onion, which are then received and absorbed by the Earth.

RS called these four cosmic spheres specific names. The Galaxy star forces he called Spirit, Solar System forces are Astral, Atmospheric forces are Etheric and Earth forces are Physical. These are the 'Above' so these are THE four players of our game. It is these four players we have to always keep in mind as they are standing behind everything else we talk about.

When we look into the inner relationships between the parts of any one of these groups, it is these four activities we identify.

This complex of activities coming from above (called the Cosmic Ether) should not be confused with the 'personal and world' Etheric activities described throughout this discussion. The World Etheric, existing within the Atmosphere only makes up one part of the Cosmic Etheric. The Cosmic Etheric also includes the World Astral and Spirit activities as well. It is a shame RS called this 'all from above' the Cosmic Ether, as it has confused many people for a long time. But he did and so we need to live with it.

We need to picture a 'layer cake'. On the outside are these World activities, which then become reflected within the layers of manifest life. Within each kingdom of nature similar activities are at play, although the way they are organised are somewhat different. The Plants for example incarnate the Physical and Etheric however they do not incarnate the Astral and Spirit activities as such. As 'bodies' they work from the outside as World activities, however they do work through the various inner activities of Etheric and Physical bodies, as we will see.

For us, Humans we have the four activities inside us, so their interactions are easier to understand. The Stars / World Spirit are the boss, albeit the Architect. They hold the 'prime intention' of any species. Inside this is the World Astral activity, carried within the activities of the various planets. These modify the forces of the Stars and provide the movement necessary within the lower layers. Inside this again is the World Etheric ' Atmosphere', were we have the Ethers and Elements as their carriers, and within that again is the Physical body. Here we have the Physical Formative Forces and the 'activity organisms'. RS outlines this interactive association in 1920 in the following. (Pg 226 17 dec 1920)

'We have within us our etheric body; it works and is active by giving rise to thoughts in our fluid organism. But what may be called the Chemical Ether continually streams in and out of our fluid organism. Thus we have an etheric organism complete in itself, consisting of Chemical Ether, Warmth-Ether, Light-Ether, Life-Ether, and in addition we find in it, in a very special sense, the Chemical Ether which streams in and out by way of the fluid organism.

The astral body which comes to expression in feeling operates through the air organism. But still another kind of Ether by which the air is permeated is connected especially with the air organism. It is the Light-Ether. Earlier conceptions of the world always emphasized this affinity of the outspreading physical air with the Light-Ether which pervades it. This Light-Ether that is borne, as it were, by the air and is related to the air even more intimately than tone, also penetrates into our air organism, and it underlies what there passes into and out of it. Thus we have our astral body which is the bearer of feeling, is

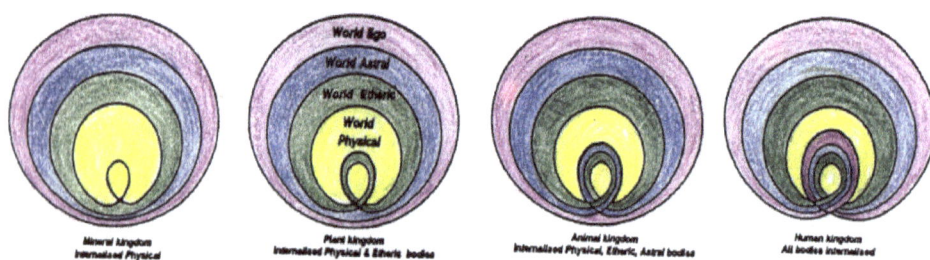

especially active in the air organism, and is in constant contact there with the Light-Ether.

And now we come to the Ego. This human Ego, (internalised Spirit) which by way of the will is active in the warmth-organism, is again connected with the outer warmth, with the instreaming and out streaming Warmth-Ether."

Ego	Will	Warmth-organism	Warmth Ether	Cosmic Forces.
Astral body	Feeling	Air organism	Light Ether.	Cosmic Matter
Etheric body	Thinking	Fluid organism	Chemical Ether	Earthly Forces
Physical Body	Manifest	Earth organism	Life Ether	Earthly Matter

Galaxy	Solar System	Atmosphere	Earth	Spheres
Spirit	**Astral**	**Etheric**	**Physical**	**Bodies**
Will	Psychology	Immunity	Body	Human
Nerve Sense	Respiratory	Circulation	Metabolic	Body Systems
Warmth	Light	Chemical	Life	Ethers
Fire	Air	Water	Earth	Elements
Hydrogen	Nitrogen	Oxygen	Carbon	Biochemistry
Cos. Forces	Cos Substance	Ter Forces	Ter. Substance	Phy. Form. Forces
Cos. Silica	Ter. Silica	Cos. Calcium	Ter. Calcium	Ca & Si
Clay	Sand	Humus	Lime	Soil
Fruit &Seed	Flower	Leaf	Root	Plant
Roundness	Pointed	Wavey	Square	Forms
Stalk	Skin	Mass	Tissues	Plant Growth
Seed	Ripeness	Size	Quality	Fruit
Germ	Seed Coat	Cotyledons	Viability	Seed
Nucleus	Mitochondria	Cytoplasm	Cell Tissues	Cell
G	A	T	C	DNA
North	West	East	South	Magnetic

If this image is expanded to take in more references the following chart arises. The four main players Spirit, Astral, Etheric and Physical activities manifest at every layer of existence, however for clarity sake RS gives the specific activities their own name. Being able to follow these associations is one of the 'tricks' of being his student.

To clarify, **manifest existence is an expression of External, World and Cosmic activities working onto Internalised activities.**

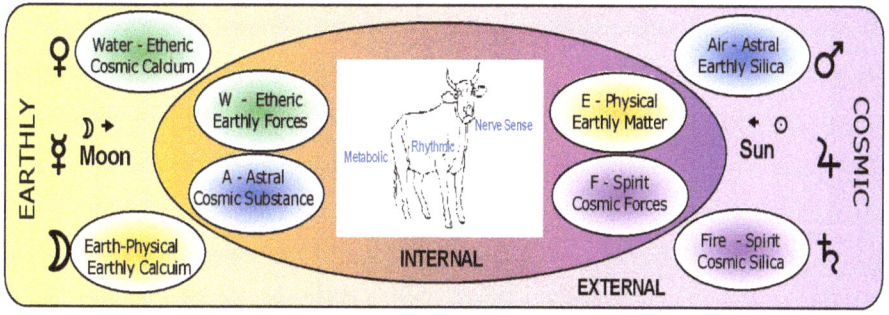

The Internal Physical Organisation

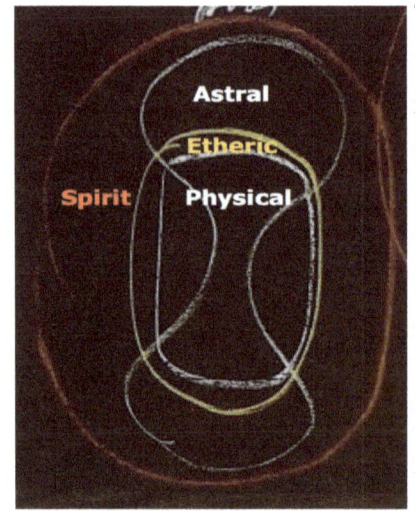

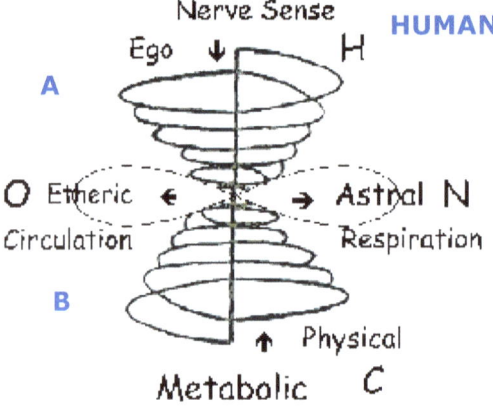

The physical organism is held together through the interaction of the four activities. The Spirit holds the 'plans', the Astral body puts the plans into action, guiding the Etheric workers, who move around the physical substances. In 1924 (Pastoral Medicine) RS provided this image to show their healthy relationship. The Physical is the white square in the middle, The Etheric is yellow outside this. The Astral is the wavey line, while the Spirit is the Red outside line.

When we come to the physical body itself RS gives more detail. Firstly RS identifies the Physical body as organising between two poles. One, that spins centripetally / catabolic, in our head region. This is where the Spirit and Astrality predominately reside (A). The other spins centrifugally / anabolic from our metabolic zone, where the physical and Etheric activities predominately reside. (B) These two processes push against each other so that in the middle in our chest region they meet and help create our breathing and circulation systems.

They do not just push against each other, they interact and intertwine with each other. The Head / nerve sense activities

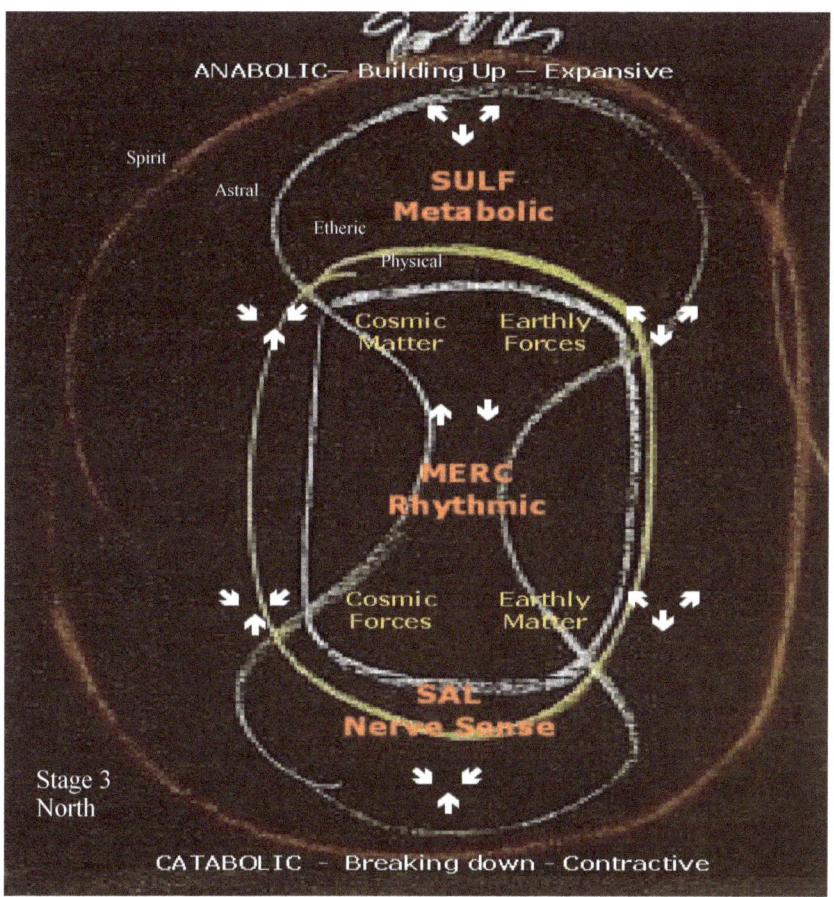

work right down into the metabolism, and provide the metabolism with a centripetal catabolic process, (C) while the metabolism works right up into the head to provide it with a centrifugal / anabolic process. (D)

In the medical lectures RS often talks of these interactions in the language of the predominate activities, so within the head he talks of Spirit and Physical processes working together, while in the metabolism he talks of the Astral and Etheric working together. In the Agriculture lectures he used the terms of the Physical Formative Forces. In the root zone there is the Spirit = Cosmic Forces and Physical = Earthly Substance. While in the top of the plant there is the Astral = Cosmic Substance, and the Etheric = Earthly Forces. This is clearly outlined in the 8th lecture, however once seen he talks of these activities throughout the whole course.

Healthy brain function is therefore an interplay between the contractive Spirit processes balancing the expansive physical / metabolic processes, coming from below. While healthy metabolic processes are maintained by the healthy interactions of the contractive Astral forces from the head interacting with the Etheric processes of the metabolic zone.

The middle Rhythmic processes depend upon both these poles to be working properly for them to then interact positively in the blood and breathing systems.

Sub Plot 1 - 4x3

If only it was this simple. There are a few sub plots.

Here is the game from 'Spiritual Science and the Art of Healing, 24 July 1924'

"*In order to understand such conditions, we must be able to look into the nature of the human being. I said that it is possible to divide the whole organisation of man into three systems: (1) the nerves-and-senses; (2) the rhythmic system (which includes all rhythmical processes); (3) metabolic-limb system. I also said that the metabolic-limb system is the polar antithesis of the system of nerves-and-senses, while the rhythmic system is the mediator between the two:* **Each of these three systems is permeated by the four members of man's being — physical body, ether body, astral body and Ego-organisation.**"

While RS talks of the predominately Spirit activities in the Head, the Astral activity is also there, but as a secondary function. So what he calls Cosmic Forces is predominately Spirit activity with a secondary 'helper' of the Astrality. Similarly, the predominately Physical activity in the head, has the Etheric as a secondary function. This 'double act' he called Earthly Substance. In the Belly region the predominately Astral activity, has the Spirit as secondary, and he called this Cosmic

Substance, while the Etheric activity in the metabolism has the Physical activity as a secondary process and he called this combination Earthly Forces.

He did this because the overall plant energetic make up does not have an incarnated Spirit and Astral activity, Hence it is confusing to start talking about the activity of the Spirit working in the root zone of the plant. There is a second image in lecture 8 that shows how the Spirit and Astrality (dominant activities) sit outside the actual physical / etheric body of the plant. So talking of them directly as 'internal processes', would be incorrect. Nevertheless the 'big' energies are active within plant processes. He uses these terms throughout the whole of the course, when talking of soil and plants, as well as when talking of animals. These four activities are the Physical Formative Forces. They have a very specific polaric relationship to each other, which is quite different to the way the Ethers work with each other.

RS generally describes the Ethers within their 'primary polarities' of Warmth and Light working from Above, with the Life and Chemical Ethers working from below upwards. The PFF work within their 'secondary polarities' (see BD Decoded) where manifestation occurs through opposites interacting, and they are THE reference most used throughout the Agriculture course, apart form the energetic bodies.

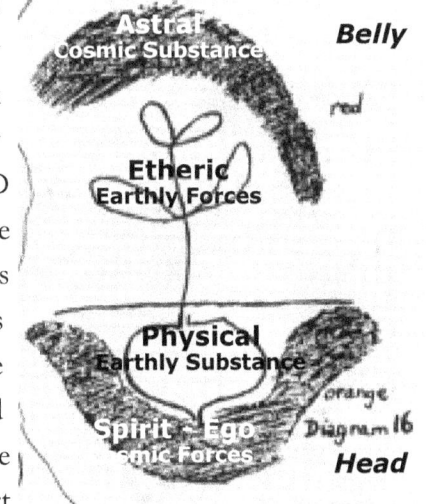

27

Sub plot 2 - The Poles

RS emphasis' that within the physical organism, the Head and the Belly regions act in a polarised manner to each other.

In the first lecture of the 1921 medical lectures he outlines how these two main polarities organise the four activities in opposite qualities. He talks of how in the Head the Physical body is 'active', which means it acts physically, while the Etheric, Astral and Spirit activities function as 'imprints' which for our purposes we can say works as Forces. In the metabolic system the Spirit acts as an imprint force, while the Physical, Etheric and Astral are 'active' and thus more physical in its workings.

This means the Astral activity in the Head is of a different quality to that working within the Belly. So if the Head Astral activity is so strong that it pushes down into the Belly, it will upset the 'normal' Astral activity that should be working there.

It is as if they are two different shades of blue. The Head Astral is dark blue, and the Metabolic Astral is light blue. The dark blue does not fulfill the need for light blue, its too much. This is the same for all the other activities as well. Metabolic Etheric is not the same thing as Nerve Sense Etheric. If these activities become active in the wrong place, illness arises.

This organisation of all 4 activities being active in both regions, all the time, is essential for RS's description of Equisetums activity with the Kidneys.

Sub Plot 3—External and Internal

This internal organisation, 4 bodies within 3 systems , sits within the external World environment, of four activities. These two realms interact with each other. External activities can enter into the internalised organism, so the World activity can replace an internal one, and it then tries to do the internal activity's job, but can not. Just as the nerve sense Astrality can not do the metabolic astral's job, neither can the World Astral or the general Astral body do the job of the internal Astral

One example of this is where physical poisonings occur. Where any chemical element becomes stronger than our body can digest, it brings with it 'unprocessed' outside forces. Eg Arsenic enhances the ways the Astrality works with regard to the Etheric body. In small doses, that can be digested, and is used to stimulate the growth of chickens, because the Astral is always needed to motivate the Etheric. However in larger doses its contractive powers takeover and it causes emotional agitation, right through to mummification, where the Etheric becomes paralysed. It is the outside excess 'World' forces doing this.

What are the Kidneys for the Human?

Karl Konig in 'Earth and Man' gives a very good overview of this. He starts by outlining the formative processes of the Kidneys during embryo development. The main features of this story are that the kidneys did not develop out of the metabolism, like the lungs and liver for example, they begin their journey from near the ears. Indeed they are the remnants of the gills in fish. An area that takes the living oxygen out of the water. "By way of their gill-breathing, the fish listen into the whole life existence, and etheric power of the surrounding water". The primordial kidneys listened to the formative influences present in the primordial atmosphere. "The music of the spheres sounded throughout all this and the kidneys heard it". Over time, the kidneys development migrated downwards into the metabolic system. A journey that still exists within embryo development.

What this means is that a 'nerve sense' activity, supported by the astral body, has found its way into the 'earthly' metabolic system. This sensory role of the kidneys carries on with their role of sensing the many balances needed within the body. As the blood moves through the kidneys, it senses the fluid and chemical needs of the body. It either strips out excesses or retains substances back into the blood flow. It does this for Urea content, water, minerals, salts and pH. It is continually 'tasting and smelling the waters' of the body. Here we see the kidneys are the organ of where the sensory astrality enters the physical body. Its proper function is dependent upon the right placement, and quality of the internal astrality.

One of the features of the relationship between the Astral and Etheric is that the active qualities of the Astral - coming from the moving planets - are needed to motivate the naturally stagnant tendencies of the 'watery' Etheric. For there to be an expansive metabolic Etheric based activity, it must be stimulated by the internalised astrality in the metabolic system. Without the metabolic Astral, and its companion the Spirit, the digestion becomes sluggish, or the Etheric can go its own way, and all manner of inflammation diseases arise.

For a healthy kidney function, this Astrally sensitive organ, must have a healthy relationship with the Etheric within the metabolic system.

What disturbs the kidneys?

There are two main kidney ailments RS discusses, an enlarged kidney, which is an inflammatory process, and the shrunken kidney. With Equisetum it is only the Shrunken Kidney, we are concerned with. Here the Internal Astral activity is suppressed or displaced from its normal relationship with the Internal Etheric, due to a few causes. Generally it is said to be the 'External Astrality', that's enters too deeply into the Kidney, so that the internal Astral activity is displaced. This breaks its relationship with the Internal Etheric within the kidney, leading to the Etheric 'becoming stagnant', and the kidney shrinking. This "External Astrality' however can take several forms.

One often mentioned source of imbalance, is bought about by an overly strong Nerve Sense system pushing into the Metabolism. This can be either or both, the Nerve Sense Internal Spirit or Astrality. This is often due to over work or over thinking everything.

RS clarifies in the latter 1921 lectures, that because of the differences in the quality of Head and Metabolic activities of the 'same player', when the internal Head Astral and Spirit work down into the Belly too strongly, they are experienced by the metabolic Astral, in the same way as if it was experiencing 'outside' World Astral and Spirit forces. Either way the result is the same.

Other passages show, that if the general Astral body enters an organ, it is also experienced as 'External'. Referencing the diagram at the top of page 25, the Astral body is identified as a 'wavey sheath', within the energetic onion. The physical systems have this basic energy specialised, internalised and refined into organs. They have differentiated this basic energy. So even the general astral sheath is too 'gross', for the internal etheric.

Another passage has the metabolic Spirit acting too strongly and also being seen as a 'World Astral' experience.

In short, the kidney has a specific relationship between the internal astral and etheric and if they are not working together properly, problems occur.

The ways the kidney Astral can be disturbed are
A) World Astral enters too deeply
B) The general Astral body can enter into the kidney
C) Nerve sense activities — Spirit, Astral and Etheric can enter too deeply
D) Metabolic Spirit either enters too deeply or not deeply enough
E) Metabolic Physical and Etheric do not accept the metabolic Astral

Quartz and Silicic acid.

The activities we have discussed so far have physical carriers. Each primary activity has several chemical agents they can work through. (see Glenological Chemistry). In the agriculture course RS mentions how the Astral and Spirit activities are carried and enhanced by Silica , while the Physical and Etheric activities are carried on the Calcium elements. We have seen how the kidneys have a particular relationship to the workings of the Astrality and so this function will depend upon the working of Silica. There are however two distinct forms of Silica. We have mineral quartz Silica often taking the form of a crystal, and then we have Silica interacting with life, through its relationship with the Etheric carrying water, in the form of Silicic acid. H_4SiO_4. This suggest that one atom of Silica can hold two molecules of water to itself. In doing so the Silica forces become more available to life processes.

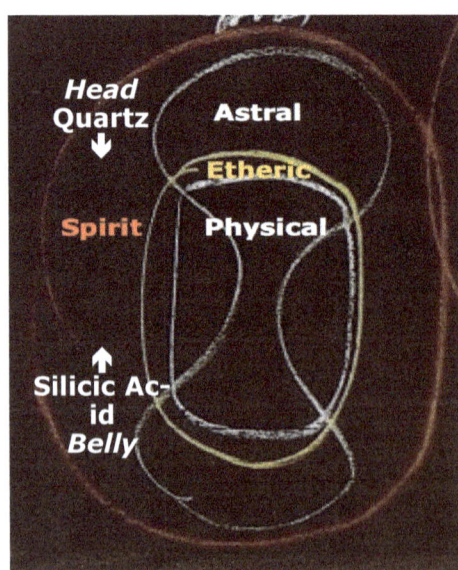

These two forms of Silica make a polarity of activity with each other. It is clear in the medical lectures that mineral Quartz in the form of a crystal, is the foundation of the nerve sense system, and using it as a remedy has the effect of pulling the Spirit and Astrality strongly inwards, from the Head downwards. It also has the effect, when used in low potencies of drawing infections to the surface, so that draining can occur, often through a 'boil bursting'.

Silicic acid, H_4SiO_4, occurs, in many places in nature, most notably though in the oceans. In this form it is found within the living processes of plants. It is drawn from the soil, and laid down mostly in the top areas of a plant, during the transpiration process. It has a more internal working action than Quartz. When used as a remedy for infection, it will stimulate the body to digest the infection from within. In this way it works with the metabolic processes rather than the nerve sense processes.

Why Equisetum is appropriate

Its phenological cornerstones are - Equisetum is a very early plant, and its ancestors where one of the three primary plant types. Grohmann places it as the middle stem formation between the 'leafy' ferns and the 'fruiting' club roots, emphasising its 'blending ' qualities. With these other two it formed the basis of many coal deposits. The main points to concern us are that it formed before the separation of the physical systems (nerve, rhythmic and metabolic) took place with the plant kingdom. This separation occurred with the flowering plants. Therefore Equisetum has the ability to harmonise the workings of the three systems together. Equisetum is essentially a stem. It has no leaves. It reproduces through spores, often from a cone like formation at the top of the shoot, however the common horsetail (Eq Arvense) forms separate shoots for producing its spores, in spring. This is then followed by the 'leaf stalks' that grow through the summer. It was the first plant to develop a vascular system, due to being forced to protect itself from drying out. Most of the plant (9/10) is below the ground and this helps to drain the soil around it. 28% of the ash is Silica, with 65% of this is silicic acid, in a opal form. Equisetum Arvense is the highest Sulphur content (4.2% of the ash) of any equisetum.

The double nature of Eq. Arvense — spring and summer shoots — offers some useful insights and actions. The spring shoots are the reproductive shoots. These are the part most filled with Sulphur, while the summer shoot is more Silica rich. F. Husemann (Equisetum, the Kidney and the Planet Venus 1993) presents a significant essay with many interesting images of this dual nature of Equisetum Arvense. The silica level in the summer shoots increases to 3 times the strength during the mid summer period, while the Sulphur levels decrease to half during this period. He outlines the various form differences between the spring and summer shoots and how they effect the astral body and concludes " We may call the tubule – related summer shoot the 'morning form', since the astral body incarnates in the limb related radiant forms. The spring shoot on the other hand, may be seen as an 'evening form' as the astral body is released through the spherical form" " The process of the spring shoot are eliminatory, as spores are released. The long lasting, radiant summer shoot...is its prime mover". A key image he gave is that 'The diseased kidney is too awake, as it were; Equisetum lets it sleep, so it may wake up revitalised". The Sulphur content allows the 'plant to sleep'.

Spores	Stems
Sulphur	Silica
Astral excarnating	Astral incarnating
night	morning
spring	summer

From this there are clear suggestions the spring 'evening shoot' is highest in Sulphur and causes the astrality to move outwards, while the summer "morning shoots' help the astrality incarnate. Given we have two forms of fungal problems one from too much astrality and the other from too little, we have a hint the Sulphur rich spring shoots will be good for dry hot fungus, while the summer shoots will be useful for the lecture 6 , high moisture fungal rots.

The Sulphur content of Equisetum is a key element in why Equisetum is so useful. Sulphur leads the Silica, which easily stays in nerve sense processes, into the metabolic system. Sulphur has a particular function. *"Sulphur proves to be a substance which plays an essential part in the reception of proteins into the domain of the human etheric body. We see that* **Sulphur does not penetrate into the astral body and the ego organisation,. It unfolds its activity in the realm of the physical and etheric body."** (Fundamentals of Therapy, pg. 60) Too much Sulphur causes giddiness and reduction in consciousness, indicting it pushes the Astral and Spirit out, it does not combine with them. The silica presence is shown in its feel, but also in the radiating spray of 'stalks' from each node, and the hexagonal scales found in the fronds.

Page 62 says *" The silicic acid has a dual function. Within, it sets a boundary to the mere processes of growth, nutrition etc., Outwardly, it closes off the mere activities of the external nature from the interior organism , so that the organism within its own domain is not obliged to continue the workings of external nature, but enabled to unfold its own activities. "*

In Equisetum, Silica is turned from a 'substance' where it works in the nerve sense realms, into a 'process', and due to the Sulphur content has a penchant to bind with the etheric body in the metabolic system. The internal astral activities are stimulated and directed to combine with the etheric and anabolic processes within the kidneys are restored.

In the 'Healing Process' — lecture 5, 15 Nov 1923 (Part 2)   RS says

"Let's sample the results of a cognitive overview that considers the human being first and then the wider world. In the method I depicted, we shift our attention from the human being to the nonhuman natural world, where we study the particular character of Equisetum arvense. We are more concerned with the process that is active in it than with the individual substances it contains. Because materialistically oriented thinking is now omnipresent, organic matter is usually described in terms of its content of protein, fat, carbohydrates, and so on. We focus on the individual components that superficial chemistry can tell us about. But the field of chemistry has changed in recent times, so now we focus on the so-called elements. The elements in an organic entity, however, are not very significant for what I have in mind. The most interesting thing to note about Equisetum is the high proportion of silica that is left behind when we analyze the plant, that is, when we separate its functions. Silica is so strongly present that it predominates and expresses its function in the Equisetum plant. Analysis reveals not so much the substance itself as its significance. Its significance is what we must recognize.

Equisetum is a plant, so we find no astral body in it; we do find a physical body and an etheric body, however. When we study Equisetum arvense, we find that silica plays a major role in it- although of course there are other plants that contain silica-while certain sulfates play a supporting role. The most important components of Equisetum, in terms of asserting their own character in the plant, are silica -not the "substance," but the silica function- and the activity of Sulphur. Then we make a very strange discovery. When we apply spiritually developed forces to understanding what is going on in the vicinity of the sulfates that are associated with silica, $SiO2$, we discover a process or a nexus of functions that we can then introduce into the human organism, either internally or, if the situation requires a different method of administration, through baths or injections. I will discuss the significance of these different methods later. Actually, it is better not to use *Equisetum* as such-although the effects are visibly present in the plant itself, they are not very permanent. In our methods of preparing remedies, we first study the functional connection between silica and sulfur, for example, and then imitate it in a medicinal preparation. We need to convert the Equisetum model into a more or less inorganic preparation that has stronger effects on the human organism than if we simply used the plant itself in the form of a tea or the like. This is the essential element in the production of our remedies.

When this functional connection between sulfur and silica is incorporated into the human organism in the right way, **it relieves the astral body in the kidney of the process it had to carry out during the illness.** That is, when sulfur and silica as they function in Equisetum arvense are introduced into the kidney, the human astral body is relieved of functions it would otherwise have to perform in the deformed kidney -"de-formed" in the broadest possible sense. For the moment, the disease process is carried out by a substitute, by a remedy that has been introduced into the body.

This is the beginning of any healing process. We must be familiar with the disease process in question, and our theory of pathological conditions must be rational. We must recognize the disease process and look for places in nature where it is copied exactly. We must not simply assume that a disease process must always be combated. Instead, we need to neutralize the process, to counterbalance it with a dynamic we recognize, such as the dynamic between sulfur and silica in Equisetum, in order **to free up the general astral body**, relieving it of functions it formerly had to perform in the diseased kidney. **We must then take care to strengthen the patient internally, through diet and so on, so that his or her inner astral forces can be applied more energetically than usual to the entire astral body.** Once we *have substituted an external function for the general astral body's excessively strong activity in the organ in question, the internal astral body, now fully normal and healthy, is able to eliminate the disease.*

This example demonstrates how we arrive at a rational concept of healing. As a general rule, healing involves substituting a process derived from the external world for the human disease process and then energizing an internal force to overcome the disease, which cannot be overcome as long as the external astral body is forced to apply its activity one-sidedly in an abnormal kidney, for example. What I have just described is or may be the case in any disease process that is due to irregularities in organs that work internally but centrifugally, if I may put it that way. Although the kidney is an organ of elimination, it initially excretes internally. If you grasp the principle I have described, you will understand that **this pathological kidney process is healed by stimulating a** Etheric *centrifugal or outwardly radiating process in the kidney by administering Equisetum arvense,* **which activates the Internal Astral activity to combine with and stimulate the Internal Etheric into action."**

How does this translate to Biodynamic language.

Because plants do not have an incarnated Astrality , there is no internal Astrality to be disrupted by the 'external' stimuli. Hence RS statement that plants can not become sick in the normal manner. In plants, the role and function of the internal Astrality is played out within the Etheric sphere, by the Light Ether, and within the PFF sphere by the Cosmic Substance. This is the silica process active within the metabolic processes of the plant. Its relationship and interaction with the Earthly Forces process is akin to the Etheric processes described for a healthy kidney. For a similar story of the Astral and Etheric interacting in the metabolism, look in lecture 4 where RS talks of what manure is, and the role of cow horns.

In RS medical examples, the 'World Astral' of some kind, causes the disturbance. This means in the case of it being the World Astral proper, the cause of fungal problems would be the third example from the medical lectures, where heat and light comes from above, somewhat similar to a pest attack.

In the case of the lecture 6 example, where RS has Equisetum providing the 'opposite process' to what he describe, he describes something very different. Yes it is a N/S soil /head invasion of the metabolic, however it is an overly active Etheric / Earthly Substance process arising from the Nerve Sense / Root, pushing upwards too strongly. Not a too strong N/S Astral push. This 'Etheric push' pushes off the 'above' processes, allowing the soil fungal processes to take residence in the leaf and fruit regions of the plant. We also need to consider how this would in turn activate the metabolic Etheric / Earthly Forces, above, which then helps to overpower the 'Internal Astral' / Cosmic Substance further. All of this suggest fungal rotting.

When this circumstance occurs in a Human — Etheric

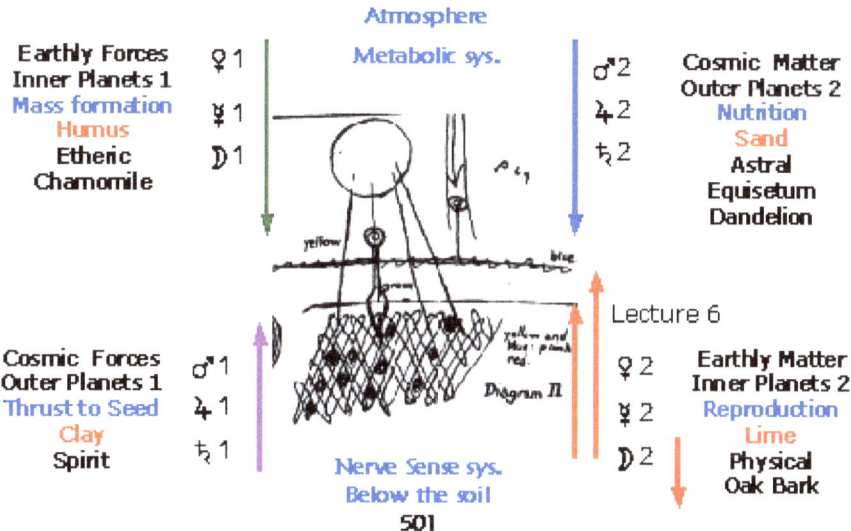

dominance, within the N/S - we would have a tendency to Hydrocephalus. If it pushed right through into the kidneys, then we would have enlarged kidneys, rather than the shrunken Kidney's given in most of RS examples. Etheric dominance in the kidneys leads to retention of water, seen as Oedema, along with a poor metabolism of nitrogen.

So these are two quite different situations. Yes it is a N/S invasion of the metabolism, but from the Etheric side rather than the Astral. RS is happy though that the Equisetum will ultimately push back the Etheric N/S processes from the metabolism. This suggestion mirrors his suggesting mineral quartz for achieving the pushing of the overly active metabolic Etheric from the Head, as in the case of migraines.

He is suggesting 'summer' Equisetum as an internal strengthener of the Cosmic Substance process, which would do two things. It would push against the upward moving Earthly Substance processes directly, and it would also encourage the Cosmic Substance to work again with the Earthly Forces, to direct its energy back into sizing things up. When controlled, the Earthly Forces would also help to regulate the Earthly Substance, it sister in the Earthly / Ca cycle.

In the lecture 6 example, not only is the Cosmic Substance processes being 'pushed off', but the upwards moving Cosmic Forces / Silica processes — described in lecture 2 — as being supported by Clay, are being suppressed, as well. This is why overly strong vitality causes fungal problems. The upward silica process, which normally carries the Calcium processes upwards with it, and through to the seed maturity, is 'swamped' by the 'vitality' inherent in the Earthly Substance / Etheric processes. This suggests that part of the solution to this problem is to also strengthen the clay process, to push upwards beyond 'the raised lake' of Earthly Substance. Homeopathic clay (D6) can be used. Clay can also be physically added to an overly sandy or peaty soil. A 20cm wide ball of clay, crumbed, per 10 square meters per year will show a marked improvement.

Let us also remember the suggestions from lecture 5 that we can also use the Oak Bark preparation (D26) to pull the overly strong Etheric activity back to the soil.

To go one step further we can reflect upon the last 2/3 of lecture 2 where RS tells of how the PFF work and what their carriers are. The Cosmic Substance is drawn in towards the Earth through Silica sand. D24 could also be used.

501 can also be used in this case as it will stimulate the light processes from above. Some care is needed not to force the plant to seed.

Hence I suggest for plant rot diseases — 501, Oak Bark, Equisetum, Clay and Sand as the basis for balancing overly strong Earthly processes.

Hot Dry

If the 'external' stimuli comes from the World Astrality, which can again be chemical toxicities, or the physical realities of too much Light and Heat, and too little water, then we have a different situation. This is the circumstance for Powdery Mildew type fungus, especially the varieties we find on Grapes. There are Powdery Mildews that appear under the 'moist' circumstances, and respond to the first remedy, cucurbits and the like, however the grapes have something else going on, that Equisetum appears to be only partially effective on.

How would Equisetum be effective in this situation?

From the earlier section it is apparent that the spring shoots are more appropriate for this type of fungus. The extra Sulphur content of these shoots work to stimulate the inner light processes, and to combine with the Etheric processes, expressed by the Earthly Forces. This combined activity in the metabolism, will push off the too strong World processes, coming from any direction. If however, Horsetail does not connect with the Earthly Forces, due to the summer form being used as a tea, it could easily cause too much astral activity in general, and make matters worse. Here we are wanting ' the spring 'evening shoot' , highest in Sulphur, to cause the astrality to move outwards'.

Hugh Courtney outlines the benefits of Lili Kolisko's suggestion that Equisetum be brewed and then fermented for a few days, until the 'rotten egg' - Hydrogen Sulphide - smell develops. The increased Sulphur activity is then utilised by applying it to the soil, rather than sprayed over the plant, to push back the upward moving Earthly Substance activity. Remember though, the Summer stems have less Sulphur than the Spring shoots, so fermenting could be a must when using them for this task. The spring shoots may not need fermenting to be effective. While fermenting them may well increase their effectiveness further.

In the example from the medical lectures, RS suggested Cinnabar, Mercury Sulphate - $HgSO4$, was the remedy, for too much World Astral.

Fungal attacks, such as Powder Mildew, can occur from the opposite direction. Human *phenomena such as diphtheria are especially able to teach us about certain subtleties in the plant organism. Such diseases should be studied more precisely, if only for the sake of discovering remedies.*

In another context I have indicated that the child's acquisition of speech is accompanied by various organic processes. While he is learning to speak, and therefore while something special is taking place in his breathing organism, something also occurs polarically in his circulatory organism, which also receives into itself the metabolic processes. I also pointed out, how what at puberty appears in a reciprocal relationship of the human being to the outer world, takes place inwardly in learning to speak. Thus this push of the astral body, which at puberty takes place from within the human being outward, takes place from below upward in the capacity for acquiring speech. (from the metabolic towards the nerve sense) *So here we have an astralizing process, and we will be able to see clearly that an interaction occurs where the respiratory and circulatory systems meet (see drawing). The astralizing process working from below upward (yellow) encounters the developing organs of speech working from above downwards (red). In this*

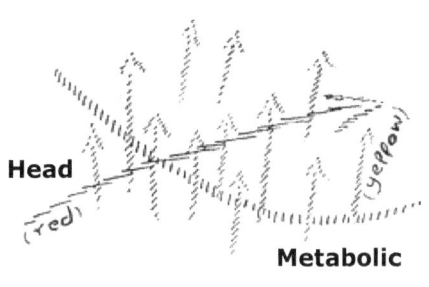

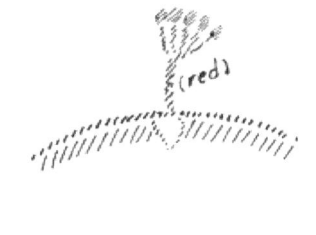

encounter the organs of speech become stronger in their capacity for speech. It is what is taking place simultaneously below in the metabolism, *that especially interests us here: this tends to work upward. The whole process is one from below upward (yellow arrows). Now, if the astrality presses upward too strongly while the child is learning to speak, we have a predisposition to diphtheric conditions. It is certainly important to pay proper attention to this.*

Let us now consider the outer earthly forces *process, we see* with plants, *that has a certain selective affinity for the process I have just described. Let this be the surface of the earth . In a plant that behaves appropriately in relation to the cosmos, the earth plays a part in the* nerve sense *formation of its roots. With growth the influence of the earth diminishes and the Cosmic Substance influence becomes stronger and stronger, unfolding especially in the blossoms (see drawing, red). What develops here is a kind of external astralizing of the blossom, which then leads to the formation of fruit. If this process, which ought to occur in the normal course of the world processes, takes place below , it can only insert itself into the water, and we have what I have just called "dysentery of the earth."*

But we can also have another situation: What takes place when a plant develops properly — the blossom unfolding always a little above the earth's surface — and can bring the Cosmic Substance right on the earth's surface (see drawing below, red). Then fungi arise; this is the basis for fungus formation, such as Powdery Mildew.

And now you will begin to guess that, if fungi arise from such a special astralizing process, the same process must take place from below upward when, as in human diphtheria, this remarkable astralization occurs in the human head. This is actually the case. Hence you find in diphtheria the tendency to fungoid formations. It is most important to consider this tendency to fungoid formations in diphtheria, and it will also show you that a truly occult process is taking place there. Everything external is really only a sign that irregular astral currents, of Cosmic Substance from the metabolism towards the head, *are prevailing within the human being.*

But when, as here, the processes work so deeply into the organism, much more will naturally be achieved by trying to find the specific remedy with which to oppose the particular process at work. One should try intermediate potencies of **cinnabar -** Mercury Sulphate. *In cinnabar we will find effects that counteract all the phenomena I have mentioned. Cinnabar expresses this even in its outer appearance. If we acquire a sound understanding of such things we will recognize that cinnabar through its vermilion color is something that in a certain way brings to expression this activity opposed to the fungoid process. That which is approaching the colorless can become fungoid. While too strong an astralization of the earth's surface plays a part in the formation of fungi, in cinnabar there is a counter-reaction to this* *astralization and thus this reddening.* Mercury a brother of Zinc is an element that strengthens the Etheric within the Physical body and so pushes off a too strong astral working. *Wherever a reddening appears in natural processes, we find a powerful counter-effect to the astralization process. You could express this in a moral formula: "The rose in blushing works against astralization." These domains of pathological-therapeutic study are really interconnected in a certain way. They guide us into this peculiar relationship of the ego and astral body to the other organs, to their laying hold of organs, to their emancipation from organs, or to manifestations of the excessive working of the astral from below upward in the human,* which for the plant is from above downwards. (2)

How would HgSO4 work. Hg in classical knowledge, (as apart from its role as a 'middle' alchemical principle) is an Etheric element. In small doses it will stimulate the Etheric body into action, and in large doses it will 'kill' the Etheric body. In my chemistry it is Zinc's big sister, and an element of the 6th ring of the Cosmic Spirit; on the Mercury 1 Arm, while on the boundary of the World Physical and the Internal Etheric. This is the arm of life's beginning. This is where algae would appear, when the Etheric internalises and picks up the first matter. Certainly an element of the Earthly Substance.

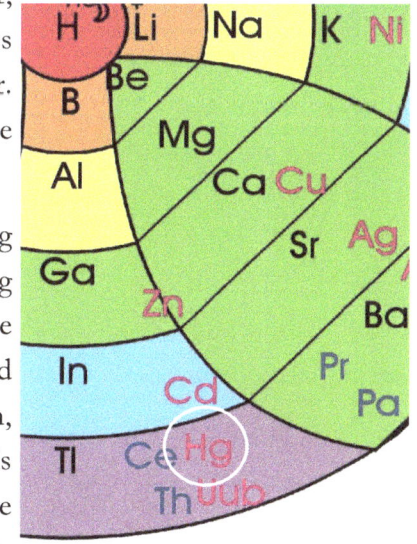

As a Cosmic Spirit ring element, it is providing guidance for how the Internal Etheric will bind into the physical organism, in the appropriate way. This will be experienced by the plant as an upwards push, from below, by the Earthly Substance, wanting the Etheric to work properly throughout all the plants growth cycle.

Sulphur is 'oily fellow', and it facilitates other elements to interact. I use it as something to make things move, especially when you want them to go to the leaf or the top areas of the plant. Oxygen of course is an Etheric stimulant and there are 4 atoms of it here.

Just as RS was pushing the 'too active soil' back into itself, in lecture 6, so here too he is pushing from the opposite direction, pushing the too strong World light and warmth, back out, by the Etheric from below.

Powdery Mildew through the Seasons

From the previous passages it is clear Powdery Mildew is a weak Internal Etheric of the Plants. One question is how do we make it stronger, to last throughout the whole growing season. Increasing Carbon in the soil via composting and mulching, correct nutrition, adequate watering throughout the season are all physical things to do.

From observing Powdery Mildew through the Seasons, I have noted that it appears often at specific periods of the year. Using the Seasonal Cycle image and Southern Hemisphere dates, the first attack is often around early August as the Spring is making its appearance, but also this is the time of the Earth's Etheric body becoming stronger above the horizon in the Southern hemisphere. Prompting the question - if powdery mildew is a weak Etheric then how is the plants Etheric activity connecting or not to the Earths Etheric. The flush of new Etheric growth coming from the Earth is not integrating with the plants Etheric, a mix of the inner planet preps and 500 at this time may help.

The next period I observed an 'attack period' was in early November when the Earth's Spirit is coming above the horizon.

This is suggesting the Earth's warmth processes are not connecting properly with the Plants Warmth organism. Is the Cosmic Forces coming from below being received

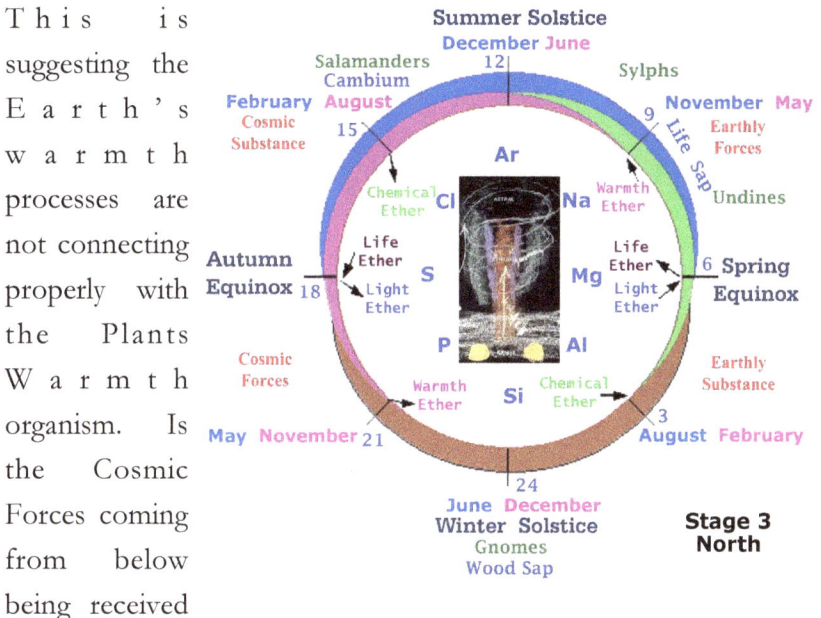

properly by the Cosmic Substance processes aided by the physical increase in warmth connecting properly. I wonder if a mix of Yarrow, Nettle and Dandelion might facilitate this combining.

The third powdery mildew period is early February, when the Earth's Etheric body is contracting below the horizon again. This suggests the plant is now left to rely completely on its own Etheric body. If this is weak, then the Cosmic processes from above take hold. Part of this solution is the strengthening of the Etheric activity throughout the whole growing season beginning in August. The inner planet preps mentioned for the August period maybe the solution here.

I have not tried these specific solutions as I have had good success with an adaption of RS HgSO4 remedy. I am still in the 'testing phase' of that remedy.

But these are my present thoughts on this matter and it seemed appropriate to share them at this time.

Clay — Aluminum Silicate

A missing link in any biodynamic discussion of fungus is the role played by clay. Lecture 2 has three direct statements about how clay allows for the Silica processes of the soil , consolidated through the winter, by sand, to be released to move upwards to the top of the plant, fulfilling the plants impulse to produce seed for the next generation. Clay achieves this mobilisation of Silica through the 'liquefying' effect of Aluminum. Aluminum's brother Boron is combined with Fluorine in glass making, to 'melt' the Silica so it can be formed into glass. Aluminum has a similar effect on Silica in nature. Boron is also used as a mineral supplement to help the sap flow run up the plant throughout the whole day. Mid day wilting is a sign of its deficiency.

This upward Silica process is what we see in springtime, coming out of the Earth as the burst of growth. This upward stream however can 'run out', or be over powered by (a) a too strong Earthly Substance activity, as described in lecture 6, and leading to a weak upward Silica stream—too little clay, or (b) from a too strong downward activity coming from too much sand in the soil or too strong inward moving external astrality due to environment conditions of dryness or excess light and heat. When the clay process is too weak, plants will grow well initially, with good leaf growth, but as soon as the plant moves off to flower fungal problems will arise, fruit will not size up and ripening is slow. The two Silica processes need to combine. The upward surge for seed maturity has to combine with the downward urge for nutritional maturity, for the Silica cycle to be complete. These naturally must also combine with the Calcium cycle. The upward Silica cycle must be strong enough to carry along the Earthly Substance and its physical nutrients with it, to provide good tissue formation. The Humus processes , both in the soil and in the metabolic processes, are needed to bring the Etheric into play. When sucked below by humus in the soil it brings life processes into the soil. But when it is above in the 'atmospheric calcium' / air moisture, it will fatten and enlarge the fruit. These four processes are needed to be working together, for large nutritious fruit, to manifest.

500 and 501

These are very interesting preparations due to the manner in which RS has 'polarised' their activity. With the cow manure preparation we see from RS descriptions in lecture 4, that cow manure is a metabolic product, of the most metabolic animal. Specifically he talks of how the etheric and astral work together in the metabolism to anchor nitrogen properly to make the best manure. The cows horn is described as a thing that reflects back these metabolic processes, when they move towards the head. These reflected back forces go to compound the strength of the metabolism we find in the cow.

Horns are NOT Antenna, as Dennis Klocek is suggesting. I have yet to see where he addresses this passage in lecture 4 - *"At these points an area is formed from which the organic formative forces* (etheric and astral activities coming from the metabolism and working forwards) *are reflected inwards in a particularly powerful way. There is no communication with the outside as in the case of the skin or hair; the horny substance blocks the way for these forces to the outside. This is why the growth of horns and claws has such a bearing upon the whole form of the animal.*

Things are quite different in the case of antlers. Here the streams of forces are not led back into the organism, but certain of them are guided for a short distance out of the organism;"

Not only is cow manure a product of the digestion, but it is a product of a supercharged digestion. So we can expect it to stimulate digestive processes. However RS chooses to place it in the soil, during the winter time—when the soil is most alive. By doing this RS is focusing this digestive ability into the soil, rather than into the top of the plant , where it might provide for good fruit sizing. Here a 'above ground' process is being bought into the Earth.

With Horn Silica we see a reverse process. Silica is the element of the Earth and the 'Head' region. It sits below all other substances and with Carbon , its brother, forms the structural basis for all life processes to develop upon. RS places this in a cow horn , as a concentrating 'organ', and into the soil during the summertime.

From RS works such as 'The Four Seasons and the Archangels' we gain some insight to his intentions. He sees the summer as the 'cosmic metabolic' time for the Earth. The center of focus of the Earth is drawn out into 'the above ground region' and it becomes light filled, during summer. By putting this 'below ground element' in the soil during the summer, he is drawing the normal nerve sense silica process into the metabolic region of the plant.

This is why he could say "the cow manure was pressing up from below, the other drawing up from above". Horn Manure is stimulating the digestive processes in the soil, enhancing the Calcium processes working upwards, while the Horn Silica has stimulated the light 'sensing' processes in the metabolism. This contracting process creates a vacuum like space, for the ground Silica processes to be drawn into. Its contracting processes are seen to be working from above. The question arises as to what would happen if we made a Winter 501 and a Summer 500, and how might we use them?

Equisetum and 501

How are they different? There is a piece given on 11 April 1921 that provides some indication. The whole lecture is well worth reading.

501   Head

500   Belly

"To be able to gain insight into the different gradations of the ego's influence in the human being, you must realize that the ego, when it wishes to act through the limbs and metabolism, is chiefly assisted by what is contained in the silica-forming process regarded as force, and that in the silica-forming process in the human head the action as substance is strongest. Thus its action as force in the head must assist the ego with diminished intensity. Now if we focus on the relationship of the human ego to the metabolic-limb system, we find the origin of human egoism in this relationship. The sexual system is indeed a part of this system of human egoism. And the ego primarily penetrates the human being with egoism indirectly through the sexual system. If you understand this, you will be able to see that there is a kind of contrast between the way the ego uses silica to work on the human being from the limb system and the way the ego works from the human head by means of silica. One could say that in the head it works without egoism. When this is studied by spiritual scientific investigation, it is possible to see this differentiation.

(red)
(yellow)

YELLOW
Head
Substance
Silica / 501
Differentiation
⬇

⬆
Blending
Silicic Acid /
Eq
Force
Metabolism
RED

If I were to represent this

remarkable activity schematically, I would have to say the following: Considering the ego as a real element of man's organization, what it does from the limb system by means of silica (see drawing, red) is essentially to encompass the human being, blending everything present in the human being in the fluids into an undifferentiated unity, so that it forms an undifferentiated, uniform whole. Then, in what is really the same process but now regarded in its activity as a force, we find the least intensive silica-forming tendency, and this works in the opposite way (yellow); it differentiates and radiates outward. From below upward the human being is held together and undifferentiated by means of silica. From above downward he is differentiated into separate components. This means that in relation to the human being the forces working organically in the head become differentiated for their work on the individual organs. In a sense they are stimulated by the silica-process belonging exclusively to the head organism to work in the appropriate way in the various organs — heart, liver, and so on. There we encounter the process which, when acting from below upward, mixes everything together in the human being, whereas when it works from above downward its action works to mold separate organs, regulating the organization through the individual organs.

We need to gain a clear conception of the results of these two tendencies in the human being — the blending tendency on the one hand and the tendency to differentiate the various organs on the other (the synthesizing-organizing activity in contrast to the differentiating-organizing activity). If we gain a clear conception of the way these two can act irregularly in a particular person, we will gradually learn to treat a person on the basis of this when something is the matter with him. "

This piece emphasis' the role Silica plays in working with the Spirit. The Spirit holds the plan that the Astral carries out, so where we have Silica being referred to as an Astral element, we must remember the Spirit is holding its hand and guiding it to action.

Here we have Silica working from the Head downwards, as Substance. Within the Metabolism Silica works as a blending Force. This provides an image for us to look at 501 and Equisetum. 501 is the Substance, albeit raised up, that brings differentiates, and Equisetum is a form of Silicic Acid - a naturally etherised Silica - that acts as a stimulator of 'inner light' within the metabolic system. It is the Force, that stimulates the Cosmic Substance and Light Ether into action, while 'blending it with the Earthly Forces. It wants things to harmonise into their natural functions for everything to run smoothly.

Horn Silica is a Nerve Sense element, however it is not simple mineral quartz. It has been matured within a metabolic period of the year , the summer. It has also been processed, so as to be more Silicic acid like. Firstly it is ground into as fine a powder as possible, and then a salt spoon full is diluted in a reasonably large quantity of water and spread over one acre of land. Especially in combination with water and stirred for an hour, how much colloidal H_4SiO_4 is formed ?

We can see it as 'moving towards' Equisetum, but not quite making it. We can see its much harsher action in the more upright growth, tendency to harden and segregate leaves, sending things to seed, that it carries the differentiating activity strongly. The Etheric has not got hold of it to the same extent as it has Equisetum. Its 'substance' upward silica strengthening role is still intact. However it also has a downward moving compressing action, that enhances nutritive and keeping quality. Here we see an enhanced metabolism, which suggests 501 has been lifted 'towards Force'. Given its role of intensifying light processes around the plant, being sprayed into the atmosphere around the plant, and being closer to the mineral, the plant would receive it as an element of the World Astral, more than the internal astral role of Equisetum. RS preparation of it has moved it closer to Equisetum, than would be normal for straight quartz, but it is not there as it has not been taken up by a plant 'living' Etheric body, as the Si in Equisetum has.

So while the Equisetum provides a downward push, it is from the inside of the plant, utilising the Cosmic Substance and Earthly Forces interaction, rather than using the 'normal' plant World Astral activity, which is used by Horn Silica, 501.

Equisetum and Bidor

In some of the examples when RS is talking of Equisetum, he takes a divergence, and talks about his Bidor remedy, and its solving of Migraines and general inflammation. This is a mineral copy of the Si, Fe and S formation of Equisetum. One easily gets the impression that Bidor is just a stronger Equisetum, but it is not. Equisetum is clearly a metabolic herb. Bidor is ground quartz combined with Iron Sulphate. It is cooked to a high temperature and ground again and taken as a 1% powder. So this is a very mineral preparation. When we look at the migraine stories, the story is of a disturbance of the metabolic zone. The Etheric has lost its connection to the Spirit and is developing too many anabolic forces. Too much uncontrolled expansion from the metabolic pushing up into the Head. The grey matter of the Head takes on too strong a digestive function and we have swelling and migraine. The solution is to use Quartz to stimulate the Astral and Spirit in the Head, into action so that

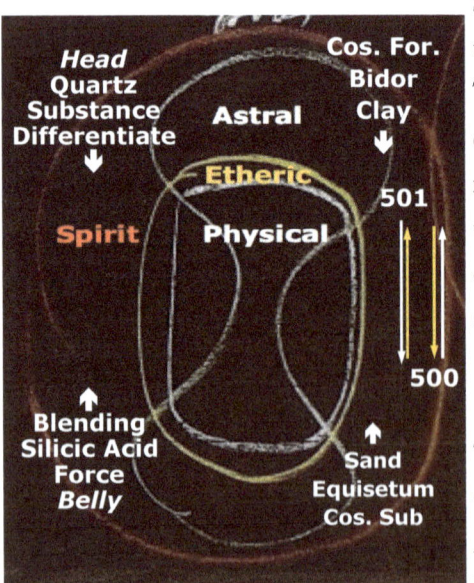

it will push downwards, pushing the metabolism back into its zone. This can be achieved just with Silica D6. RS shows how the S is used to settle the metabolism, and then for Fe to act as a mediating harmonising element to help the two systems work together again. So while he may have taken his lead from Equisetum, he has used its fundamental action, and changed it from the Silicic acid pole to the Quartz pole by making it from minerals, to effect the opposite outcome. He is pushing from the Head into the Belly, not the Belly to the Head. This is similar to the primary action of 501, only achieved by combining Silica with Sulphur rather than putting Silica in a horn through the summer.

Weleda considers this remedy to be one of the 'odoron' remedies, whose main aim as a remedy 'type', is to achieve healing by balancing all the systems. So they emphasis all three minerals activity, however the lecture descriptions place an emphasis on the Silica activity being dominant.

I have not done any plant trials with Bidor yet, but combined with Cinnabar and Spring Equisetum we may have something very special in the 'fight' against hot dry fungal problems and various insect pests. My trials with cinnabar on Grapes have shown encouraging results.

Potency Choice

We have a hint in lecture 6 of the 1921 lectures about the choice of potencies. *"you can conclude that the system most similar to outer nature is the metabolic-limb system. If something is lacking there, you must use the lowest potencies (1-10). As soon as you have to deal with the middle system, you need intermediate potencies (11-20). When you have to work with the head, when something has to do with the spiritual in the head, you have to work with the highest potencies (21-30)."*

Equisetum's Planetary Ruler

In Part 2 of this article, I will look further into this topic, as there are suggestions that Equisetum is ruled by either Saturn or the Comets. The basis of these suggestions need to be addressed and this is not the place to do that.

As a summary of what is presented here the indications that stand out indicating a planetary association is (a) Eq is predominantly a stem > Rhythmic processes , which work towards harmonising above and below (b) being from the Old Moon period before plant differentiation > a harmonising ability of all the activities (c) Working via the Silicic Acid pole of the metabolism > a metabolic strengthener (e) Having a strong Sulphur activity > an element that works predominantly with the Etheric, while helping other elements work into life processes (f) Harmonising the Internal Astral with the Internal Etheric > which occurs in the metabolism (g) works on the kidneys, who perform a 'sensing' harmonising function > which are ruled by Venus and Libra.

Within Biodynamics we have three planetary organization's we can reference.

A

	♄ Saturn	**Spirit**
Cosmic Silica	♃ Jupiter	
		Astral
	♂ Mars	
		Equisetum
	♀ Venus	
		Etheric
Earthly Calcium	☿ Mercury	
		Physical
	☽ Moon	

The first is the order outlined in the first lecture of the Agriculture Course, where we have the Cosmic Outer Planets sitting in opposition to the Earthly Inner Planets. The Silica processes carry the Spirit and Astral

56

into manifestation, while the Inner Planets carry the Physical and Etheric into manifestation. From the text we see the Silica processes active in Equisetum, however they are bound by life processes, and work to harmonise the Internal Astral with the Internal Etheric activity in the Kidney. The Sulphur influence further connects it to the Etheric sphere. There is a polarity image here, where Quartz and Phosphorus work with incarnating the Spirit into matter, while Silicic Acid and Sulphur, help to incarnate the Astral into the Etheric. Hence Equisetum sits in between the Mars and Venus activities.

The second planetary reference available to us, as a development of the first, when it is combined with the more complex image of RS understanding of how the four activities work in the three physical organization's, as shown below. In lecture 7 & 8 in the 1921 medical lectures RS talks of the

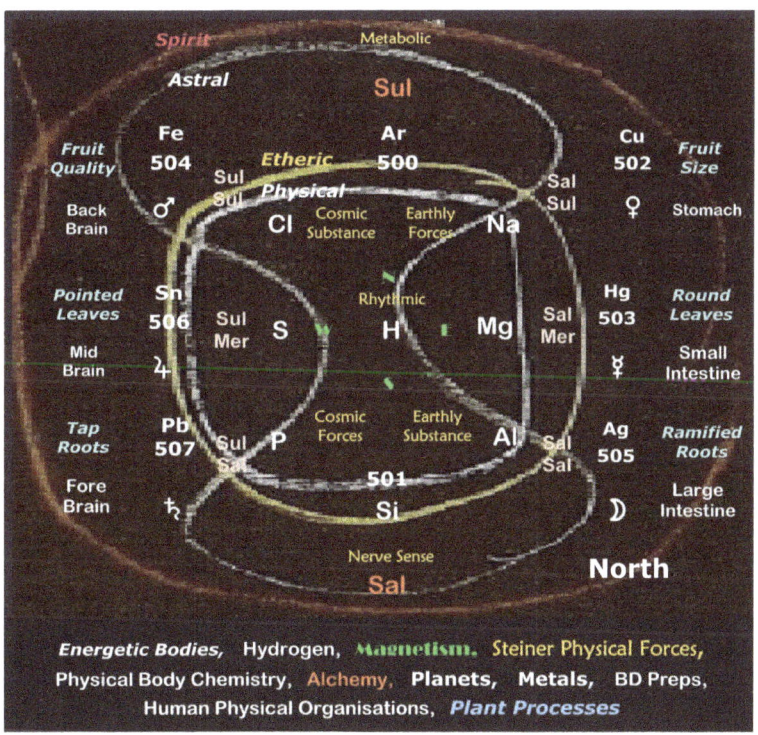

planetary activities and the metals. When placing these in relationship to the alchemical processes he had described in the earlier lectures, one was challenged to the ask the question - '**How can the three alchemical processes become the six planetary activities? In short how do we threefold the threefold**?' This is an exciting process that is described in my 'Alchemical Chemistry' article. Ultimately though all of these indications can be summarised by this diagram.

Following on from the thoughts of the previous references. Equisetum is working to combine the Earthly Forces and Cosmic Substances within the metabolism. Again as an interplay of the Mars and Venus activities that predominantly supports the metabolism, to work inwardly against an overly strong nerve sense pole coming from the Soil.

This diagram also provides an image of how all the other Biodynamic preparations sit with regards this overall form.

This basic planetary ordering is taken three steps further when we see Lecture 2 and Lecture 6's planetary indications working as a team into the building and manifesting processes

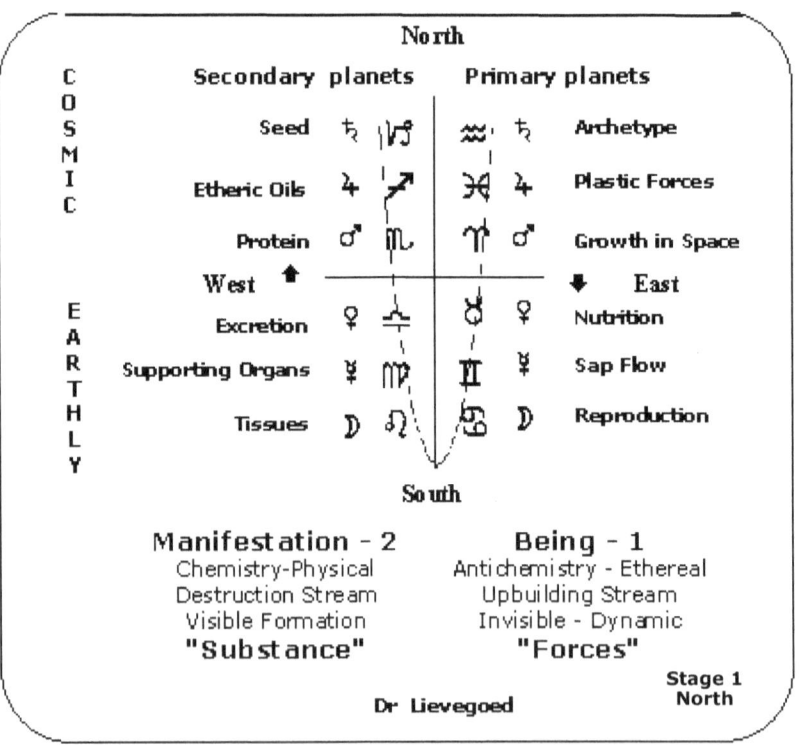

of creation. Dr Lievegoed alerts first alters us to the first Stage of this ordering in his 1951 lecture / booklet. This has the planets as a double process, with one side being the incarnation processes, or coming into being side of the planets, While the second side is the excarnating, or manifesting, processes of the planets. This order finds its basis in the planetary rulership's of the Zodiac Constellations. It is a more complex reference, however it does make sense of RS's planetary indications in the Agriculture course. Being 12 fold, we are looking at a Spirit World story. The previous 7 fold planetary story is a Astral World story.

Staying with the theme developed above, we have Equisetum being a plant of the Mars Venus interaction. In the double planetary order each planet has two functions, one as a process of pulling things together so something can manifest and the second being a physical activity of an organ or biochemical process. While Lievegoed's language is somewhat difficult, his intentions can be seen and experienced. The incarnating process begins with Saturn 1, where the Spiritual impulse of a species is taken up, at Jupiter 1 this archetypal plan is adapted to the reality at hand eg a particular environmental condition. At Mars

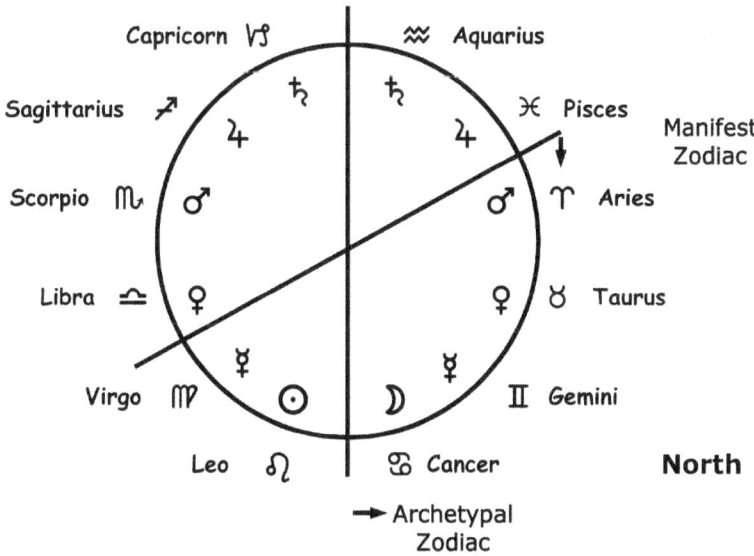

1 the Astral forces are accumulated so that the primary intention is moved forward. At Venus 1 the Etheric environment is prepared , by the necessary living resources needed for life, be drawn together. Mercury 1 strengthens the Etheric living processes and implementing the plan into the prototype stage of development. Moon 1 brings together the physical resources needed for the 'item' to become manifest , with 'pre marketing testing' being carried out. Moon 2 is the 'birth' or germination of all this accumulated activity. This is the new plant growing through the cotyledons or the newborn child, through until they walk. Then Mercury 2, the leaves and stems are extending in prolific growth and the nutrition of the plant becomes self sustaining, just as the intestines now 'feed' the body. The product is 'shopped around' for its market . Venus 2, sees the plant readying itself for 'reproduction' by coming into flower. The human is able to discern what is good for itself and what is not, through the kidney function, while the project finds its supporters and friends. With Mars 2, the Astrality enters and proteins are formed as pollen and fertilisation is achieved. The Gall Bladder produces bile and proteins can be digested, while projects start to achieve their own momentum. Jupiter 2 sees the Spirit work with the Astrality to have more complex oils and tannins form, while the fruit now swells. Here the liver function drives blood purification and nutrition, while providing other gastric juices for the digestion of fats. Projects experience success. Saturn 2 is where the Seed reaches maturity, and consolidates the new archetype into the seed ready for the next crop. The Spleen as the Saturn organ regulates how the rhythms of our inner selves with those of the outside world. It looks for foreign things in our blood and removes them. It stands us in relationship to the World. For a project this is its completion. It has achieved its goal and now the 'what next' has to be contemplated.

Lievegoed emphasis' how these individual activities can be seen to work with their polarity. So Saturn 1 and Moon 2 have a relationship. Jupiter 2 will work with Mercury 1 and so on.

With Equisetum we can see that it is helping the Astrality find its place within the life processes, indicating Mars 1, with the ultimate outcome being that the normal Kidney function is restored Venus 2. So within this more complex reference we can say Equisetum is a herb that supports the Mars 1 / Venus 2 dynamic.

Epilogue

This essay is as much an exercise in 'finding Equisetum' as it is an example of what Dr Steiner's medical work has to offer to Biodynamic Agriculture. Firstly it provides a much larger, stronger and more reliable knowledge platform for Biodynamic Agriculture to answer its questions from. Secondly it provides a treasure trove of remedies we could utilise for the many problems we face. A few medical doctors have helped us in the past. Drs Konig and Lievegoed come to mind, and no doubt there could be many more.

Modern Biodynamics needs to make a shift from its simplistic 'Threefold and Ethers' story, to the more 'adult' medical approach, where all fur activities are found in the three physical systems. The existing storyline does not lead to significant problem solving, or understanding of the Agriculture Lectures. The course will never make sense in this 'context'. Too many apparent contradictions remain. The medical lectures provide the Agriculture lectures with the right context for them to be understood and applied.

There are week long International Professional Medical Trainings (IPMT) being offered annually, in many parts of the world, by the Anthroposophical Medical Group from Switzerland. They are willing to accept serious Biodynamic students to attend. Dr Steiner's medical work offers Biodynamics a pathway out of the terrible malaise it finds itself in, where 'Any Old Story Will Do', and 'organics plus the preps' seems our only horizon looking forward. We are better than that. We have enormous potential to bring conscious spiritual science into a practical maintenance of nature. Rather than be 'the salt and pepper for organics' we could be its innovators, direction givers and problem solvers, as Dr Steiner was.

Appendix 1

Spiritual Science and the Art of Healing
lecture 1 - 17th July 1924

Spiritual Science and the Art of Healing provided at the RS archive, is a different translation of the same lectures that appear in the book 'The Healing Process' from which these passages are from.

Degeneration and Regeneration

Thus we learn to see that the human being as a whole exists in a polarity of opposites. In any organism where generation and regeneration take place, degeneration must also be present. Degeneration and regeneration are in constant flux in any organ we look at, whether liver, lungs, or heart. Isn't it a strange figure of speech, for example, when we say "The Rhine is flowing there"~ What is the Rhine? When we look at it, we usually have the flowing water in mind, not the riverbed. But the flowing water is never the same from one moment to the next, although the Rhine has been there for hundreds and thousands of years. What is the same at any given moment? The changing flow! Similarly, everything inside us is in a constant state of flux between degeneration and regeneration. Degeneration / centripetal provides a vehicle for the spiritual element. In every normal human life, centripetal degeneration and centrifugal regeneration are in balance, and our real soul-spiritual capacities develop in this state of balance. The balance can be disturbed, however, when an organ grows rampantly because it fails to counter its unique degree of regeneration with enough degeneration. Or the opposite can occur: an organ fails to counter a normal degree of degeneration with enough regeneration. A physiological phenomenon becomes pathological, and the organ atrophies and dries up.

We need to comprehend this state of balance in order to understand how it is disturbed by excessive degeneration or regeneration. Having understood such disturbances, however, we can widen our view to include the whole wide world and discover the external natural process that can counteract disturbed degeneration or regeneration. For example, if we consider first a human organ that is disturbed as a result of excessive degeneration and then turn our attention, sharpened by spiritual scientific insight, to a plant as it appears in nature, we can recognize the regenerative process in that particular plant. It becomes evident that certain species of plants always contain regenerative forces that correspond exactly to the regenerative forces in human organs. According to the general law I have just explained, we can discover that regenerative forces are always present in the human kidneys.

Let's assume that these forces are too weak and are overwhelmed by degenerative forces. Looking around in the plant world, we perceive that the common scouring rush, **Equisetum arvense, contains regenerative forces that correspond exactly to those in our own kidneys. We prepare a remedy from Equisetum; in the appropriate way, through the** circulation and through digestion, we bring it into contact with the organ it is intended to affect. This remedy strengthens the kidney's weakened regenerative forces. Similar remedies can be developed for other organs. Once we have acquired the basic insight, we can use forces we find in the world outside ourselves to reestablish the balance between degeneration and regeneration in human organs. Whenever we discern regenerative forces that are too strong and degenerative forces that are too weak, whether in the kidneys or in other organs, **we** *must strengthen the regenerative process with remedies derived from primitive fernlike plants."*

The trick of this passage is that regeneration comes from the Etheric, yet Equisetum is an astral stimulant. The connection of it to pushing out the 'outside force' and the moving onto stimulating the internal Etheric activity, is not clarified.

Spiritual Science and the Art of Healing,
lecture 2 - 21 July 1924

Getting back to our diagnosis of reduced sensory capacity in the kidneys, we need to introduce the appropriate silica process. I described how one aspect of respiration combines oxygen with silicon and distributes it throughout the body. We must make this activity move more strongly in the direction of the kidneys. To do so, we must know how to come to the aid of an organism that cannot produce enough silica for the kidneys. Somewhere in the external natural world, we must recognize the process corresponding to what is lacking in the organs in question-in this case, the kidneys. How do we find ways and means of guiding the silica process into the kidneys in particular?

We discover that kidney function, especially inasmuch as it is also a sensory function, depends on the human astral body. which is the basis for the specific degenerative processes we call elimination. Therefore, we must introduce silica from outside and stimulate the internal astral body to guide it into the kidneys. We need a remedy that stimulates the silica process in the kidney in particular. When we look for such a remedy in the plant world, we discover Equisetum arvense, the field scouring rush. One characteristic of this plant is that it contains a great deal of silica. If we were to administer pure silica, it would not get to the kidneys. Equisetum also contains sulphates, however, which affect the rhythmic system and the organs of elimination, especially the kidneys. When sulphates are as closely associated with silica as they are in Equisetum arvense, they smooth the way for silica to reach the *kidneys.*

This individual example demonstrates the actual pathological activity in kidney disease. Our method of formulating the appropriate remedy was very precise; we looked for a substance to replace the missing process in the kidney. Step by clearly understandable step, we bridged the gap between diagnosis and treatment.

Course for Young Doctors — lecture 1
2 Jan 1924
Ego too strong in the Kidneys

You can only get a conception of the possibility of illness when you know that what constitutes illness when it takes place in the liver, may be healthy when it takes place in the heart and so on. For if the human organism, working from out of the I organization, could not bring forth the warmth that must be present in the region of the heart, the organism would, for example, be unable to think or to feel. But if these same forces were to invade the liver or kidneys it becomes necessary to drive them out again, to put them back, as it were, within their original boundaries. Now, in external nature there are substances and activities of substances which can take over, in the case of every organ, the activity of the etheric body, of the astral body, of the I-organization. Suppose the I organization is taking too strong a hold of the kidneys. By giving equisetum arvense in a certain way, you enable the kidneys to do what the I-organization is doing in this abnormal, pathological condition. In this pathological condition, the I-organization is taking hold of the kidneys but in the way that ought only to happen in the heart, not in the kidneys. Something is going on in the kidneys, which ought not to be there, but which is there because the I-organization is pouring in its activity too intensely. We only get rid of this condition if we introduce artificially into the kidneys an astral *activity which is an equivalent of this activity of the I-organization. That is what you can introduce into* the kidneys if you really succeed in making equisetum arvense active in the kidneys. The kidneys have a great affinity with equisetum arvense. The activity of this substance throws itself into the kidneys, and the I- organization is sent out. And when the I-organization is given back to its own tasks it has a curative influence upon the diseased organ. You can call up the higher bodies, so-called, into health-giving activity when you drive them out of the diseased organ and set them again at their own proper tasks. "then, through a reactionary force which arises, these higher bodies can actually work curatively upon the diseased organ.

Spiritual Science and the Art of Healing
lecture 5   15 Nov 1923   (Part 1)
Physical and Etheric resisting the Astral

"Just as latent warmth is released and transformed into real warmth that is expressed in physical effects, the ether body, astral body, and I are expressed in the physical human being. We understand the human being only by looking at how these four members interact.

To gain an idea of this interaction, let's consider a specific detail, such as the human kidneys and their function. In every limb and organ of the human being, the four members of our human constitution work together to a greater or lesser extent. What we see when we study the kidneys by examining a corpse or making other physical observations is only the sum total of physical effects. These physical effects, however, are pervaded and energized by what I called the etheric body--specifically, by the part of the ether body that includes the vital functions of the kidneys. The ether body, in turn, is pervaded by the astral body. Only the interaction between these members enables us to understand the makeup of the human being, even in a single organ or organ system.

Let's take a case of irregular kidney function. Since you are all experts, I do not need to go into detail. When we understand the *entire issue from the perspective of anthroposophical research. we realize that the irregularity (*too strong nerve sense activity*) causes the kidney's physical and etheric functioning to resist the astral kidney function in some way. This is a typical case. Astral kidney function, which we can perceive only when we have emptied our consciousness, is resisted by the physical and etheric organization of the kidneys. When such resistance occurs in any living organ, its astral organization* (External Astrality) *must intervene much more thoroughly and energetically than it normally would, or otherwise the organ would atrophy. The result at least in specific cases, and the cases I describe are always concrete – is an intensification of the part of the External astral organization that corresponds to the kidney and its activity. In other words, astral kidney function becomes much stronger than it ought to be, and the kidney places much greater demands on the astral body than it ought to in the overall constitution of the human being. From the perspective of anthroposophical research, the astral body performs this work in the kidney by withdrawing activity from the rest of the human body. The* external *astral process in the kidney actually should not be there. Excessive demands are being placed on the astral kidney because of specific abnormal developments in the physical and etheric kidney.*

At this point in the diagnostic process, we must know that the External astral portion of the kidney is active in a way that is not necessary when the organism is functioning normally. In the kidney's present pathological condition, the physical or etheric kidney places extra demands on the External astral component, which responds with activity that is not needed in a healthy kidney. Here we have the first element in understanding the nature of the patient's illness. To a thinking person, disease processes should be a great riddle, because they are actually natural processes. But normal processes are also natural processes. How do the abnormal processes of disease get mixed up with normal ones. As long as we see the human being as an arbitrary network of physical substances and functions, we have no basis for distinguishing physiological phenomena from pathological ones. We can make this distinction, however, when we know that the kidney can metamorphose when physical processes develop in it that do not occur in a normal kidney whose physical, etheric, and astral components are in harmony. This is our first insight.

Now the question is, how can we combat this disease process, which is simply due to excessive demands being placed on a supersensible portion of the human constitution? How can we make the Internal astral being function normally again? I want to keep my explanations concrete and detailed. I will not discuss serious kidney disease because the same principles are evident in milder illnesses. But simply in order to suggest how to treat such a kidney, I would like to take a specific example as my starting point. We know, to begin with, that we must **relieve the external astral body of its work in a kidney** that has been "de-formed" in the broadest possible sense. In this kidney, the human astral body is doing something it should not be doing. We must remove the External astral body from this abnormal kidney process

reference needed
Internal Astral working into the Internal Etheric

If diagnosis reveals what is going on in a diseased kidney, the process that must be applied therapeutically is the same, but on a different level. For example, the disease process may be counterbalanced by giving the patient a combination of sulfur and silica as a substitute for the particular pathological process I have recognized. A cure is accomplished through a treatment, implemented by the astral body, that imitates the disease process on a different level. For example, if I introduce the equisetum function into the human organism, it stays in the ether body, and the external astral body is relieved of its work in the diseased kidney.

Thus diagnosis and treatment, which exist in parallel today and can come together only on a purely empirical basis, become a unity. In recognizing the nature of a disease process in this way, we discover, for example, that Equisetum arvense imitates a specific kidney process.

Spiritual Science and the Art of Healing
16 Nov 1923

In this same way you can also see how the silica I mentioned yesterday works in Equisetum. I told you that Equisetum incorporates silica in a particular way that influences kidney function. Modern anatomical and physiological studies fail to consider that the sensory-nervous system can be separated from the circulatory and metabolic systems only on an abstract level. In a certain sense, all organs are sense organs. The kidney is a particularly important abdominal organ in this respect. **Using silica as it occurs in Equisetum in the way I indicated yesterday increases the kidney's sensitivity** *and positively affects processes in the human organism that stem from a deadening of the kidney's inner sensing ability,* caused by a weak internal astral activity.

Course for Young Doctors
Pg 213   24 apr 1924

Silica acts in the same way that the human kidney does. If I give the silica which is in equisetum to a patient I build up the phantom of a kidney in his renal region. The phantom then replaces the internal astral activity at this place. This presses out old kidney substances and permits some new kidney substances to form from what is in flux, just as it forms there in any case after seven to eight years. However, one accelerates the process by producing this phantom. One should realize that an organ and the activity which forms it are always present together, and that this activity always rigidifies into the organ. This is where you encounter the fluidic human being.

Course for Young Doctors
pg143 22 apr 1924

Upon what does healing depend? It depends upon knowing which substances, which forces must be applied to the human being in order that the process of disease may pass over into the healthy process. Such knowledge is transmitted, for instance, by the fact that one knows: Equisetum, within the human organism, takes over the activity of the kidneys. When, therefore, the activity of the kidneys is not sufficiently cared for by the internal astral body, I shall see that they are cared for by equisetum. I give support to the astral body by means of equisetum arvense. Here for the first time is the answer to what is really happening. The same process in the external world which leads to equisetum also takes its course in the human kidneys. The equisetum process must be studied in connection with the kidneys. This leads us to the domain of healing.

Equisetum and Migraine

The Healing Process
2.Sep.23

Here again, my concern was to discover parallels between the process summed up in the symptoms of migraine and an activity in outer nature. On the one hand, we have the migraine syndrome and on the other an opposite process that we see in the activation of silica by sulfates in Equisetum arvense.¹ As you may know, Equisetum arvense contains approximately 90 percent silica. Tomorrow we will speak about silica's very significant functions with regard to the sensory-nervous system and everything related to it. In Equisetum arvense, however, silica interacts with sulfates, forming a loose combination that can be recreated in manufacturing by using a resinous binder.

Simply call to mind an image of Equisetum arvense. See how stiffly it develops, how the formative process of silica is allowed to predominate throughout, how the plant holds back its growth so that it does not flower (flowering occurs in connection with normal metabolic processes). A true inner view of both processes-the one that is expressed in migraine symptoms and the wonderful process that takes place between silica and sulfates in Equisetum arvense-immediately suggests that these are polar opposites.

This does not mean that direct use of Equisetum arvense is helpful against migraine. At this juncture we become clearly aware of a peculiar circumstance-that, although certain vegetative processes in the human organism are similar to plant processes in some respects, they are radically different in others. The point, therefore, is not simply to take the activity that occurs in Equisetum arvense and incorporate it into the human organism, but rather to "animalize" it first, so to speak.

This animalization can be accomplished in the laboratory by imitating the plant process in an inwardly active way, using silica on the one hand and sulfur on the other. Sulfur is the actual active ingredient in Equisetum arvense and can be used as is. Using binders that play subordinate roles, the sulfur is combined with silica by introducing the iron process into the production of the remedy, animalizing the entire Equisetum process. The result is a preparation whose efficacy is based on how it is produced. How this therapeutic preparation is manufactured makes it clear that it represents a process involving silica, iron, and sulfur. The resulting preparation-or rather the activity that has simply been brought to rest in the preparation-is reactivated and set in motion when it is introduced into the human digestive system and used in treating migraines. As I said before, our doctors call it Biodoron. Almost without exception, this migraine remedy has proved to be extremely effective.

Spiritual Science and the Art of Healing

lecture 3   24 July 1924

Migraine

Now let's contemplate an example that shows how therapeutic insight results from understanding the pathological workings of the human organism. Before we discuss this example, however, a few preliminary comments are in order. A remedy of sorts, a remedy that we all need, is always present in the human body itself. I am not using the term "remedy" very precisely here, but you will immediately understand what I mean. In human beings, the I-being and the astral body always tend to descend too deeply into the physical body and ether body. We would prefer to perceive the outer world dimly rather than clearly, to rest rather than to be active. Our natural preference for rest is a constant illness that must be cured. We are healthy only when our bodies are continually healed by the iron in our blood, which is present for that purpose. This metal prevents the astral body and I from uniting too strongly with the physical body and ether body. Iron therapy is an ongoing internal cure. Whenever there is too little iron in our blood, we become listless and want to rest, and whenever there is an excess of iron, we experience involuntary activity and restlessness. Iron regulates the relationship between the physical and etheric bodies, on the one hand, and the astral body and I, on the other. When this relationship is disturbed in any way, increasing or reducing the body's iron content will restore the correct proportion.

Now let's consider a form of illness that is not held in great esteem by the medical establishment, and understandably so. This illness is confusing because the contributing factors are not readily apparent, and a plethora of remedies are available. It seems that every drug manufacturer has developed a remedy. This illness, which does not elicit a great deal of respect from the medical establishment, but is very unpleasant for the people who suffer from it, is migraine. Migraine seems confusing because it really is a very complicated illness.

In studying the human head, we are struck by the central location and wonderful network of the extensions of the sensory nerves. The design of the internal portions of the sensory nerves, located toward the middle of the brain in the human head, is really quite miraculous. This nerve structure is the most highly perfected part of our physical form, because it is where the effect of the human I on the physical body is most pronounced. In the way our sensory nerves make their way into the body, interconnect, and bring about an inner division of sorts in the entire body, the organization of the human body transcends that of animals to a very great extent. This structure is a wonder. The human I-being, the highest member of our human constitution, has to intervene to regulate this miraculous structure, and it is very easy for the I to fail from time to time, leaving our physical constitution to its own devices. The I often may not be strong enough to pervade and organize the so-called white matter of the brain thoroughly. When the physical and etheric bodies slip out of the I, something resembling a foreign structure is incorporated into the human body.

The white matter of the brain, as you know, is surrounded by gray matter, which is much less finely structured. Ordinary physiology sees the brain's gray matter as more significant than the white. This view is inaccurate, because gray matter is much more closely associated with the process of nourishing the brain, of accumulating substances, while white matter, which is located in the brain's center, is supported by the spiritual element to a much greater extent. All parts of the human body are related, however, and each member affects every other member. As soon as the I begins to withdraw from the brain's central white matter, its gray matter succumbs to disorder. The astral and ether bodies are no longer able to take hold of the gray matter correctly, and irregularities develop throughout the interior of the head. The 1-being tends to withdraw from the central part of the brain and the astral body from the brain's periphery. The entire functioning of the human head shifts. The central part of the brain begins to serve conceptual activity to a lesser extent, becoming more similar to gray matter and developing digestive activity of a sort that should not occur, while the gray matter itself becomes more of a digestive organ than it is meant to be and secretes too actively. Throughout the brain, foreign bodies are incorporated, and excessive secretions accumulate. All of this reorganization in the head, however, works back on the more subtle respiratory processes and especially on the rhythmic activity of blood circulation. The entire human body is in a significant, although not very profound, state of disorder.

At this point, it is important to ask how we can reincorporate the I into the nervous system, specifically into the extensions of nerves toward the interior. How do we force the I back into the central portion, of the brain that it has abandoned? We achieve this by administering, silica, a substance whose mode of action I described in the first two lectures. If we use pure silica, however, the I is driven back into the central area of the head's sensory-nervous system, but the surrounding gray matter remains unchanged. To regulate the digestive process in the gray matter simultaneously, so that it does not overflow and is rhythmically incorporated into the entire normal context of our human constitution, we must also administer iron, whose ongoing role in the human body is to regulate the connection between the rhythmic system and the entire spiritual system.[1]

We immediately notice that the cerebrum, in particular, tends to develop irregularities in its digestive processes. Nothing happens anywhere in the human body, however, without affecting other parts. Consequently, subtle disorders develop in the entire digestive system. If we again study the connection between an outer substance and the human body, we find that sulfur and its compounds regulate digestive processes throughout the body.

We have now mentioned three perspectives that need to be considered when dealing with migraine: using sulfur to regulate digestive processes whose irregularities become most apparent in the cerebrum, using silica to induce the I to control sensory-nervous activity, and using iron to adjust disturbances in the rhythms of the circulatory system. These three perspectives result in true insight into the human body and clarify the entire migraine process, which-as I said-conventional medicine tends to disdain. The body itself seems to be telling us to manufacture a remedy composed of silica, sulfur, and iron. The result is the migraine remedy, based on anthroposophical research, that is now being distributed worldwide. This remedy is extraordinarily effective in regulating the I-being so that it intervenes in the body in the right way, in counteracting disturbed rhythms of blood circulation, and in promoting appropriate digestive activity that radiates throughout the human body.

Those who understand the human organism know that migraine is ultimately only a symptom of the fact that the ether body, astral body, and I are not working within the physical body in the right way. This same phenomenon can cause other disturbances in the organism. It is not surprising, therefore, that our migraine remedy is suited to regulating the interactions of the I, astral body, ether body, and physical organization on a more general level. Whenever these members are not working together properly, our migraine remedy-which is more than just a migraine remedy-will offer relief We call it a migraine remedy simply because it works on a phenomenon whose most radical symptoms appear in migraine. I have used this particular remedy to clarify how we study the nature of an illness according to anthroposophical principles and how we produce the remedy once we know the effects of substances on individual members of the human constitution.

The Healing Process
16.11.23

The illness I'm talking about is migraine. To understand it, we need to know that it is caused by an activity that does not belong in the head, a hypertrophied version of the subtle metabolic activity that normally occurs in the head. We need to relieve the head of metabolic activity that does not belong there. How is this done~ First of all, we must administer a substance that will take over and perform this metabolic process. After what I said earlier, you will see that this substance is silica. Silica must permeate the sensory system that is irritated in migraine, where it relieves the head of the pathological migraine process. But first we must make sure that silica's activity reaches the head. If we want the preparation to be taken orally, we must make sure that it gets to where it is going and does not become stuck somewhere in the digestive system along the way.

To accomplish this, we must make the astral body as active as possible, so that it carries the silica upward in ascending waves through the digestive process into the head. We can promote the upward flow of silica only by making the astral body as effective as possible. This means that all factors hindering the vital working of the astral body must be eliminated from the circulatory system, which mediates between the abdomen and the head. Sulfur does this, so our remedy must contain both silica and sulfur, suitably processed. The remedy, however, is delivered via the rhythmic system, which must work both upward and downward in the human body. As you know, we can follow the upward and downward rhythms of respiration and circulation. These rhythms are best supported by the inherent activity of iron. If we intend something to flow upward but do not want it to get trapped, or if we want it to remain there without making demands on the entire body, we need to produce a preparation containing iron, sulfur, and silica, prepared in a specific way. The resulting remedy, Biodoron, relieves the head of the migraine and then reincorporates that activity into the total human body in the right way.

Hayfever the opposite of the Equisetum function
Reference needed

Other disease processes demonstrate the polar opposite of what I have just described. Once again, I will not use the example of a serious illness to introduce the general principle, but rather one that, although it is extremely unpleasant for the patient, attracts relatively little attention compared with deeper-seated illnesses. This disease is hay fever, or allergic rhinitis. In attempting to combat hay fever, we must realize that we are dealing with a fundamental constitutional disorder based on a peripheral reduction in the forces of the third, inwardly mobile human being, the astral body. The origin of hay fever can be traced back to childhood, when generalized and generally disregarded disorders appear that then become specialized, appearing later in life as hay fever. If we know that **hay fever involves a decline in certain astral functions that prevent the astral body from reaching the physical and etheric bodies,** *our first concern must be to energize the astral body internally and redirect it to what it should be doing.*

In a pathological condition such as hay fever, more externally directed centrifugal effects are apparent and must be actively counteracted. In kidney disease, we counterbalanced or offset the illness; we saw that the astral body simply needed to be freed from its abnormal work in the diseased kidney and then energized and strengthened in order to work in the direction of health. This is not true of processes such as hay fever. In such cases, we cannot begin by offsetting the disease process. Instead, it must be counteracted by a comparable process working in the opposite direction. We have found that the activity an astral body has stopped performing because it no longer has access to the physical and etheric bodies can be restored by using the juices of certain fruits. The fruit must have rinds, and centripetal effects must be evident within the fruit itself. A preparation made from the juices of such fruits is administered in ointment form in milder cases or as injections in more severe cases. We have already experienced considerable success with this treatment, which drives the astral body back into the physical and etheric bodies. Dr. Wegman has successfully treated many patients with our injectable hay fever remedy.

Our way of thinking does indeed yield ways of energizing a sluggish astral body. Depending on the specific fruit that is chosen, the processes these injections stimulate have an affinity for particular organs. We need to investigate which organs are affected and what tendencies reveal the affinities. The injection-induced processes demonstrate that physical symptoms due to a sluggish astral body can indeed be corrected by offsetting the astral body itself. In the previous example, we neutralized the disease process, but in this case we counterbalance a process in the area we hope to affect. In choosing a treatment, we must distinguish between centrifugal therapeutic processes, such those I described for the kidneys, and centripetal processes, as in our hay fever remedy, for example. On first consideration, such remedies may seem to have been dreamed up out of nothing, and in fact that is what most of our contemporaries will believe. Consequently, I have placed great emphasis not only on the need to produce such remedies but also on the need to implement our own school of medical thought in our institutes. Confirming the efficacy of our remedies, however, is not the same as testing remedies developed through empirical methods applied on the physical level. In the latter case, we depend heavily on statistics, which tell us whether the remedy is helpful in a high proportion of cases. If we begin by applying a method such as the one I have discussed here, however, our clear understanding of a specific disease process reveals the cure, so diagnosis and treatment become one and the same thing. The situation is this: If diagnosis reveals what is going on in a diseased kidney, the process that must be applied therapeutically is the same, but on a different level. For example, the disease process may be counterbalanced by giving the patient a combination of sulfur and silica as a substitute for the particular pathological process I have recognized. A cure is accomplished through a treatment, implemented by the astral body, that imitates the disease process on a different level. For example, if I introduce the Equisetum function into the human organism, it stays in the ether body, and the astral body is relieved of its work in the diseased kidney.

Thus diagnosis and treatment, which exist in parallel today and can come together only on a purely empirical basis, become a unity. In recognizing the nature of a disease process in this way, we discover, for example, that Equisetum Arvense imitates a specific kidney process. If we recognize that the inner character of gall excretion in certain illnesses is the same as the process we find in Cichorium intybus, the plant process shows us how to relieve the astral body of the gall-excreting function it would otherwise have to perform in the liver. We make advances in healing when diagnosis is no different from treatment, while treatment becomes a truly rational science.

The Three Worlds in Biodynamics
Glen Atkinson   14.3.25
RS Italics

This essay has been developed over many years. Interestingly I missed the importance of the Three Worlds reality when I skimmed through Theosophy in the late 1970s. I noted the evolutionary story, as being very similar to the Hindu stories I had come across and moved on.

Over the years I kept coming across a threefold enfolding process in various parts of creation, but had not 'croked' that it was RS's Three Worlds story sitting behind it all.

This reality is the basis of RS Nature Stories, even though he has not stated this clearly. Nevertheless, If we want to understand how Dr Lievegoed's 1951 planetary suggestions begin to solve the 'Great Planetary Contradiction' in the Agriculture Course, and then move through to the Seasonal Stories, with our sanity intact, then this is the pathway. The Agriculture Course does make sense and it can be understood, it just needs putting in its right place in the Three Worlds story. It is the story of the Second , Soul / Astral World. The activities behind Manifestation that can help form substance.

This all helps clarify the Manifest World order, which is different to both Lievegoed and the Agriculture Course planetary orders, yet vitally important when working with plants and animals through the yearly seasonal cycle.

This 'moving of patterns' from one order to another is very common in Astrology, where an archetypal reality such as 12 Constellations, can be picked up and applied to the Seasons, beginning at the Spring Equinox of the Northern Hemisphere, and we call them the Signs of the Zodiac. They are not the same thing however the 12 fold vibe of the spirit is present in both. Astrology is full of such games, so doing this to Biodynamic astrology seems a natural thing to do.

The Three Worlds

One of Rudolf Steiner's fundamental stories is told in his book Theosophy, in 1904. It is described at the RS archive as *"Theosophy begins by describing the threefold nature of the human being: the body, or sense-world; the soul, or inner world; and the spirit, or universal world of cosmic archetypes. A profound discussion of reincarnation and karma follows, concluding with a description of the soul's journey through regions of the supersensible world after death. The book closes with an outline of the path to higher knowledge."*

This is the story of how these Three Worlds manifest for the human. Given they sit as the basis for Manifestation, we could expect they are of significant value for working with all the other kingdoms of nature. In Biodynamics is concerned with the Soil, Plants and Animals, which begs we ask **How BD works with the Three Worlds?**

RS calls these Worlds the Spirit World, Soul World and Manifest World, and he tells us that they exist one inside the other, in the same time and space, but function as three individual energetic fields, with their own inner laws of working. Referencing

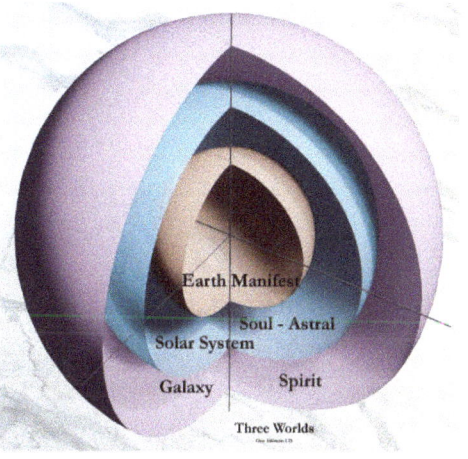

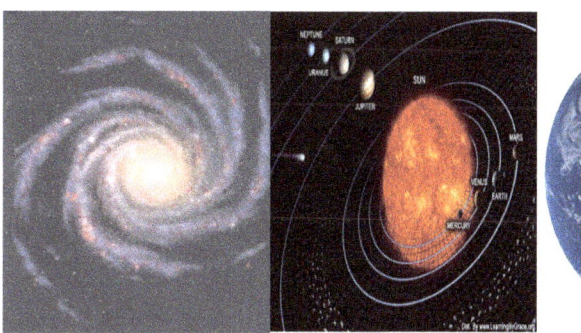

his other lectures it is fair to conclude he is describing the astronomical fact that we exist on a planet, in a Solar System that exists within a Galaxy. These are three real electro magnetic energetic dimensions, within each other. As fanciful as the idea of the Three energetic Worlds might appear, it is indeed an astronomical fact of Physics.

Once this three layered view of reality is identified, RS's lectures take on an increased degree of complexity as references to them pop up here and there. They are one of those things one passes over until you see it. These connections bring some clarity, as he often describes more about the inner laws of each World.

In this quote from 9 October 1920, RS is quick to show that the three physical organisations of the Nerve Sense System, Rhythmic system and Metabolic system are an expression of these 3 primary Worlds.

"We find above all that when through Imagination and Inspiration we enter the spiritual world in full consciousness, it immediately appears to us to be threefold. Hence we can speak of the world, and of our theme, the evolution of the world and of man, only when we have come to the point we have now reached. Only now can I speak of how a man, confronted by the external world, by all that manifests itself to the senses, is really facing the spiritual world in its threefold nature – facing actually three worlds. Once the veil has been lifted which creates the chaos, we no longer have one world only before us, but three worlds, and each of the three has its definite connection with the human being.

When we succeed in penetrating this veil of chaos – later I shall be showing how we can also describe this as crossing the threshold of the spiritual world – we perceive the three worlds. The first of the three is really the world we have just left, somewhat transformed but still there for spiritual existence. When the veil of chaos has been thrust aside, this world appears as though it were a memory. We have passed over into the spiritual world; and just as here we remember certain things, so in the spiritual world we remember what constitutes the physical world of the senses. Here, then, is the first of the three worlds.

The second world we encounter is the one I have called in my book, Theosophy, the soul-world.
And the third world, the highest of the three, is the true spiritual world, the world of the spirit.

To begin with, I shall give you only a schematic account of all this, but from the way these three worlds are related to man you will gather many things about them. To these three worlds as they appear in three ascending stages – the lowest, the middle one, and the highest – I will then relate man's three members – the head; then the breast-organisation embracing all that is rhythmical, the breathing system and blood circulation; thirdly, the metabolic-limb system, which includes nutrition, digestion and the distribution throughout the body of the products of digestion, all of which engender movement. All this has to do with the metabolic-limb system."

This reference gives us the beginning - the Three Worlds - and the end of the story. The three Astronomical spheres are the beginning and the Physical organisation are the end. In between these two there is a lot more that can be said about how these activities show up in the other kingdoms.

A question that arises is **how do the Three Worlds unfold into its later forms?**

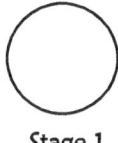 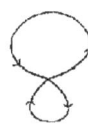 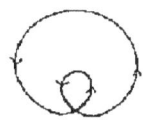

Stage 1	Stage 2	Stage 3
Archetype	Forming	Manifestation
Thesis	Antithesis	Synthesis
Passive	Pulsating	Enfolded
Spirit	Soul	Manifest
The Field	Around the Form	Body
Galaxy	Solar System	Earth

We can look to the most basic life process of cell division for a hint. The cell goes through 3 phases. It is happy as a single cell, it then begins to pulsate within itself, DNA divides, and the cell separates into two cells. Three distinct stage. Stability, Movement, Differentiation.

We have to start with the Spirit World, the expanse of the Stars. These are the primary generators of energy, particles and movement in our Creation. They are the Creators of Archetypal forms through their constant beaming of forces outwards for millions of years. Given there are Billions of Stars all beaming force at us from all directions we have to accept sciences assertion that we are at the apparent center of a holographic field of force, which is dynamic and forever changing due to the multitude of astronomical phenomena taking place within this field. Luckily astronomy is relatively predictable, with many recurring cycles. Even luckier is that Humans have been observing these cycles for many thousands of years. Creation is not chaotic, indeed it is quite predictable.

Around the Stars are formed the Planets, who are the 'beings' of the Astral Soul World. They form in the magnetic rings found within the Sun's magnetic field. They suck up the material particles floating about from the Sun and other Supernova. They move in constant orbits around the Sun, bringing predictable movements and changing tones to the activity of the Solar System, as their spheres warp from the planets gravity. Later the Planets internalize as the Astral Body and their constant movement keeps the Internal Etheric body active and

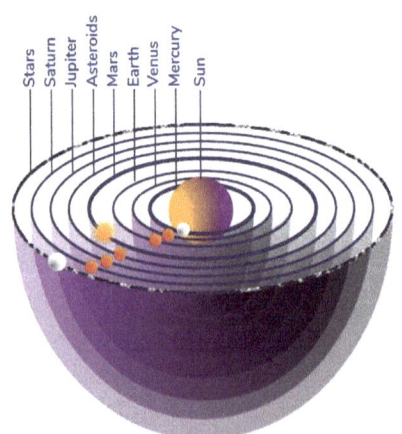

healthy, keeping everything alive. Oh but **what is this Etheric that I speak of ?**

The Earth is an accumulator of physical particles and brings the activity of all the forces into appearance. But creation does not stop with the consolidation of particles of matter. These particles interact with each as positively and negatively cations and anions in 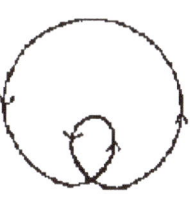 the activity of Chemistry. Chemical elements combine and release according to physical laws of chemistry and electro magnetism, until one day around 4 billion years ago, single celled organisms manifested in an anaerobic ocean. This cellular life continued until one day a energy producing mitochondria bacteria joined with it, in a process called endosymbiosis, around 1.5 to 2 billion years ago. Giving the subsequent organism its own internal powerhouse. A third stage occurred, around the same time, when these two joined with a photosynthesizing organism to allow for a self contained oxygen producing organism. Blue Green algae. The oxygen released by them firstly 'rusted' the surface of the Earth and then it began to accumulate within the magnetic 'Atmosphere' of the Earth. This act of releasing oxygen was the first step in the development of **the Fourth World — the Cosmic Etheric**, which exists in the 20% oxygen content, from just below the Earth surface to the edge of our present Atmosphere. This 'Life Sphere' has been 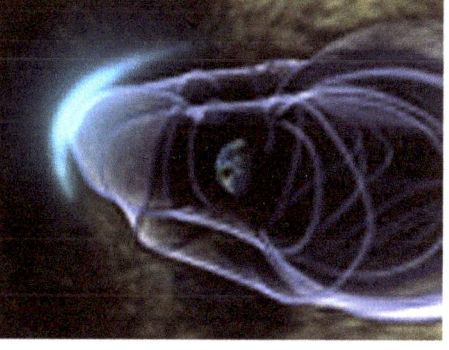 developed by the Life processes of the Earth. A second substance that anchors the Etheric into life processes is **Calcium**. It too is a product of Life. All of the solid Calcium on

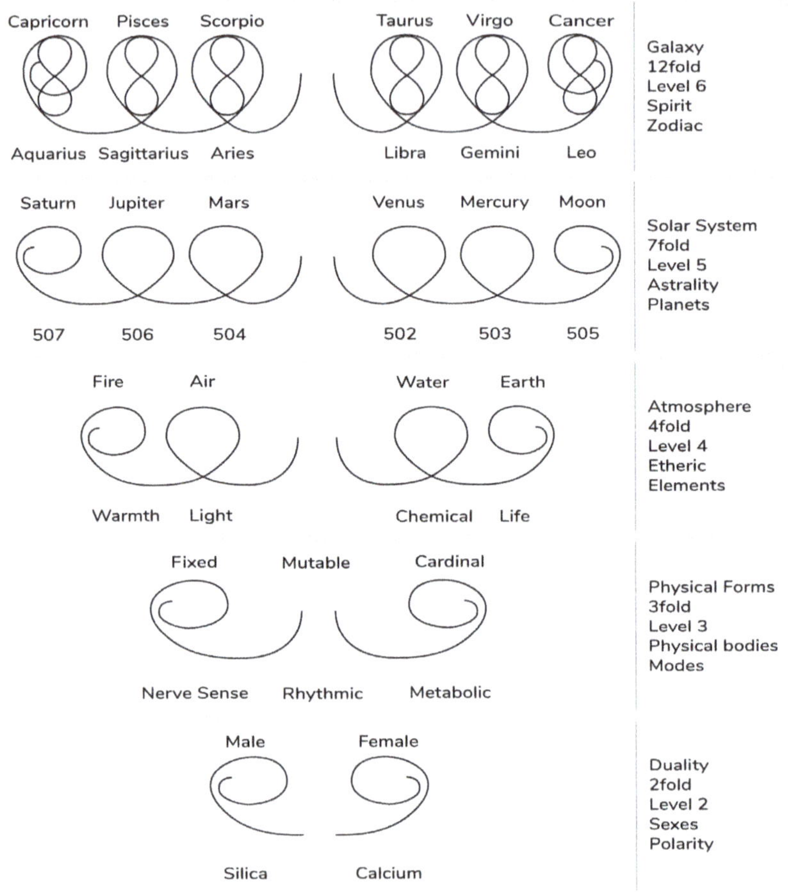

Capricorn
Pisces
Scorpio
Taurus
Virgo
Cancer
Aquarius
Sagittarius
Aries
Libra
Gemini
Leo
Galaxy
12fold
Level 6
Spirit
Zodiac
Saturn
Jupiter
Mars
Venus
Mercury
Moon
507
506
504
502
503
505
Solar System
7fold
Level 5
Astrality
Planets
Fire
Air
Water
Earth
Warmth
Light
Chemical
Life
Atmosphere
4fold
Level 4
Etheric
Elements
Fixed
Mutable
Cardinal
Nerve Sense
Rhythmic
Metabolic
Physical Forms
3fold
Level 3
Physical bodies
Modes
Male
Female
Silica
Calcium
Duality
2fold
Level 2
Sexes
Polarity

BIODYNAMIC VORTEX (b)

Galaxy
Solar System
Atmosphere
Bodies
Sexes
Earth
Duality
Physical Ring
Etheric Ring
Astral Ring
Spirit Ring

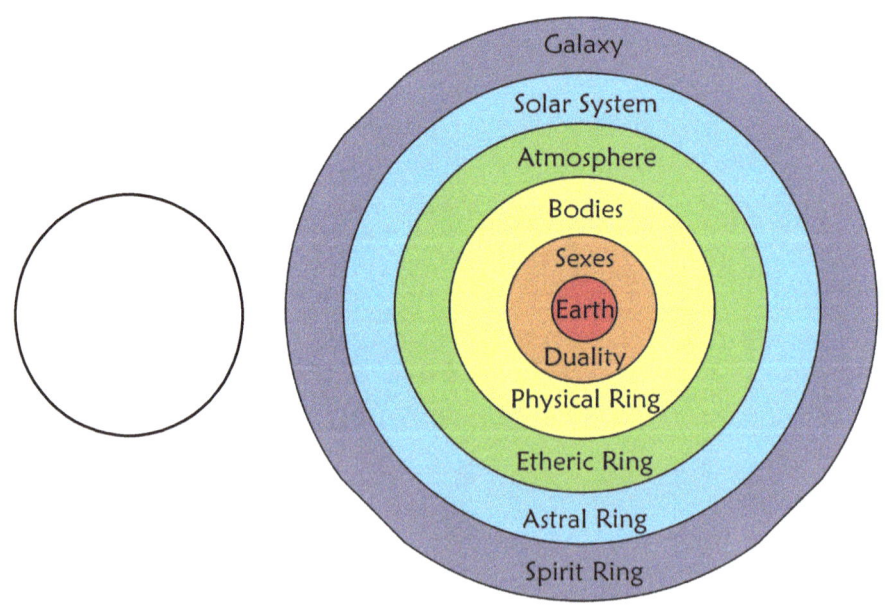

the planet is a residue of coral, seashells or bones. As wild that may seem, this is an accepted scientific fact. Life has produced its own environment within which it can multiply.

While this is called the **Fourth World,** there are two other stages in this story that need to be clarified. One is the **chemical dynamic sphere** from which physical forms arise. Firstly this is a Chemical process of positively charged cations and negatively charged anions, that form complex molecules. This sphere of polarity eventually shows in the phenomena of the sexes, which are needed by the majority of Life to multiply and bring forth physical bodies, with their three fold organisation..

This next stage, the **physical bodies form**, initially is anaerobic microbes, but later as more complex lifeforms develop they clearly display the threefold organization. They act as external male and female agents that come together to produce a physical offspring. Twofold becomes Threefold. We have a realm of general Physical Bodies created. Level 3. This is the pool of DNA we inherit from the Earth.

This substance is bought to life by the Etheric. We come again to the fourth World of the Cosmic Etheric activities. As the Cosmic Etheric developed, with the release of oxygen and calcium so the complexity of lifeforms could develop.

This means within our external creation, up to the Galaxy, we live within a 6 layered game of chess', where any move on any layer ripples through all six layers. This means in practical terms we do not have to just work with the physical material layer of creation, we can also intervene at the other energetic layers of organization. We just need to know what the rules of each organization are.

RS talked of this organization in his many discourses of the 2 fold, 3 fold, 4 fold, 7 fold and 12 fold activities, at the base of various bits of creation. This organization and numbering, based upon Astronomy is the same as used by Astrology for millennia. As this is ALL THERE IS. This is what you will see when you look. These 6 'rings' are called the Cosmic Activities, within this whole story of 'What is There', as these are the spheres or Rings of the External energetic dimensional realities in Space. This organisation is the first Stage, it is Spirit World story. There are two more structural Stages to come The Soul World and the Manifest World.

The Spirit World of the Stars is the realm of the Archetypes, or base plans of creation. This is because the stars beam a constant vibration through space for millions of years at a time, and so create standing waves patterns as they move through other energetic fields. These are the source vibes that later become species of physical bodies.

When we look at the organisation of the Constellations in the Spirit World, we can see how the planets organise at this stage. Dr B Lievegoed gives an interpretation of this pattern in his 1951 lectures (2) . He identifies the double sided nature of each of the planets, and when seen as a cycle there is an inward moving 'incarnating' half and an outward moving 'excarnating' half of the cycle. He then talks of how the various planets polarize with another. Saturn 1 with Moon 2, Venus 1 with Mars 2 and so on. This polarization is a hint of the planetary Soul World activity we will come to next.

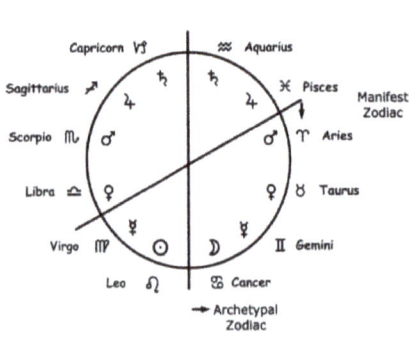

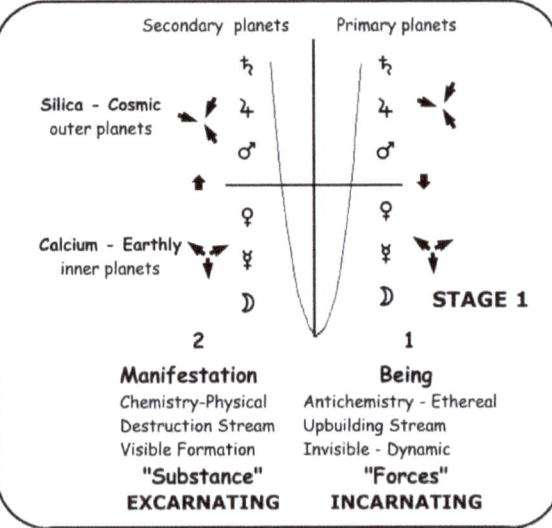

As we are passing by this constellational image, there are a few interesting sights to see. You may note that Lievegoed's story of Incarnation and Excarnation, when extended to the Constellations produces a Aquarius to Capricorn zodiac, moving in a clockwise direction. Being Saturn based, I use this 'journey' as a clarifier of ones purpose through to achieving its outcome.

Also note the Aries to Pisces zodiac moving in a clockwise direction. I call this the Manifest Zodiac. We can ask way does this zodiac begin with a Mars energy?

Then we can observe the Cancer to Leo Zodiac, moving in an anti clockwise direction. I call this the Archetypal Zodiac, as it is involved in the large scale Precession of the Equinoxes story of human evolution. Eugen Kolisko showed how the animal kingdom unfolds according to it, and it is useful in Chemistry as well, helping to clarify the activity of the Rare Earth elements. This is a significant zodiac. In some ways more important than the Aries to Pisces Zodiac.

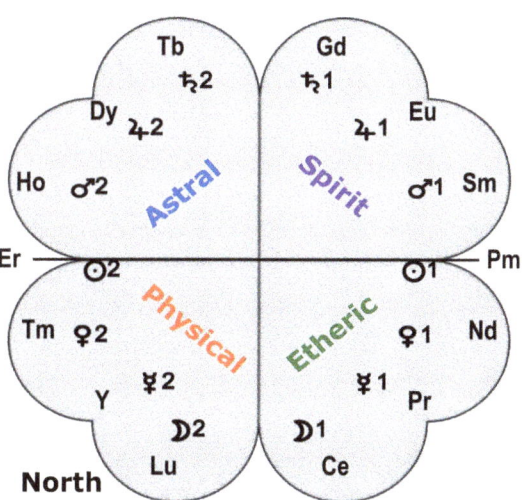

Astral Soul World

The planetary spheres, encountered by distant star forces, are themselves the result of one particular stars activity. We need to bring **movement** into our story. Things are not stagnant as the Rings diagram suggests, everything is moving very fast. When matter moves it electro magnetically polarizes and creation begins unfolding according to the Golden Mean mathematics. The Cosmic activities polarize and form electronic gyroscopic spheres, where the Spirit and Physical activities form a vertical axis and the Soul Astral and Etheric activities form the

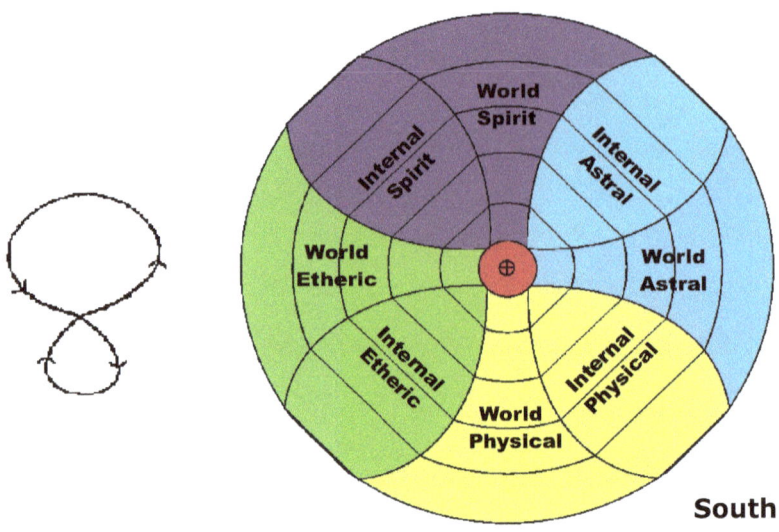

horizontal plane. Due to the Stars spinning it also forms a magnetic onion, upon which the planets accumulate the cosmic dust that is squirted out along the horizontal plane of the Sun's field. The planets are indeed compost heaps of collected matter. These spherical planetary spheres, within the Suns field warp as the planets move, so their resonant tone created as a star force moves through it, is forever changing. These planetary tones are the Soul World activity caught by the distant stars standing wave and carried into Life as a sub species formative influence.

The cross section image has the World Activities on the primary cross (vortex) with the secondary cross (petals) being the place where the Internal bodies manifest. Here we have a polarity forming of the External World and Internal World. As the external world moves so does the Internal Bodies adapt.

For how the planets organize in the Soul World we can pick up Lievegoed's thread, where he talked of the various processes in Life forming due to a polarized interplay of the various planets. He pictured them as opposites, however in RS Agriculture Course, he presents a more detailed description. It takes a little piecing together however once seen it is hard to not see it. We must bring in the role of movement, and this is imaged in the formation of

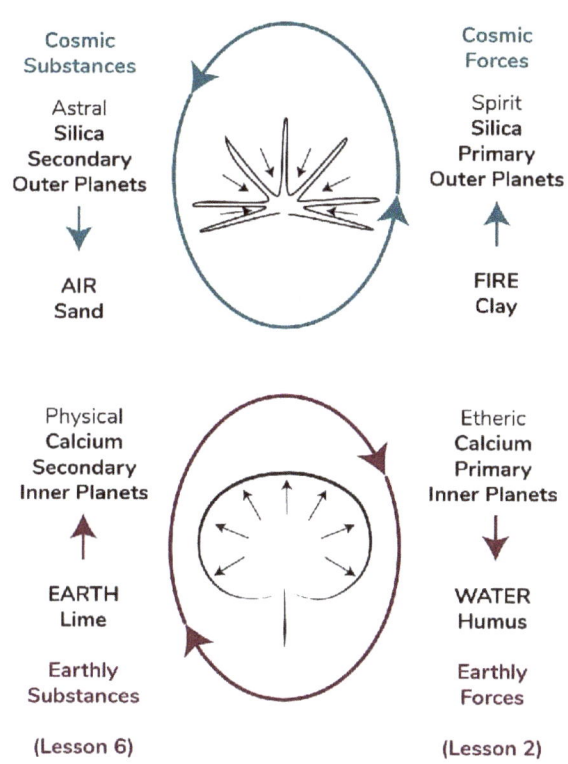

(Lesson 6) (Lesson 2)

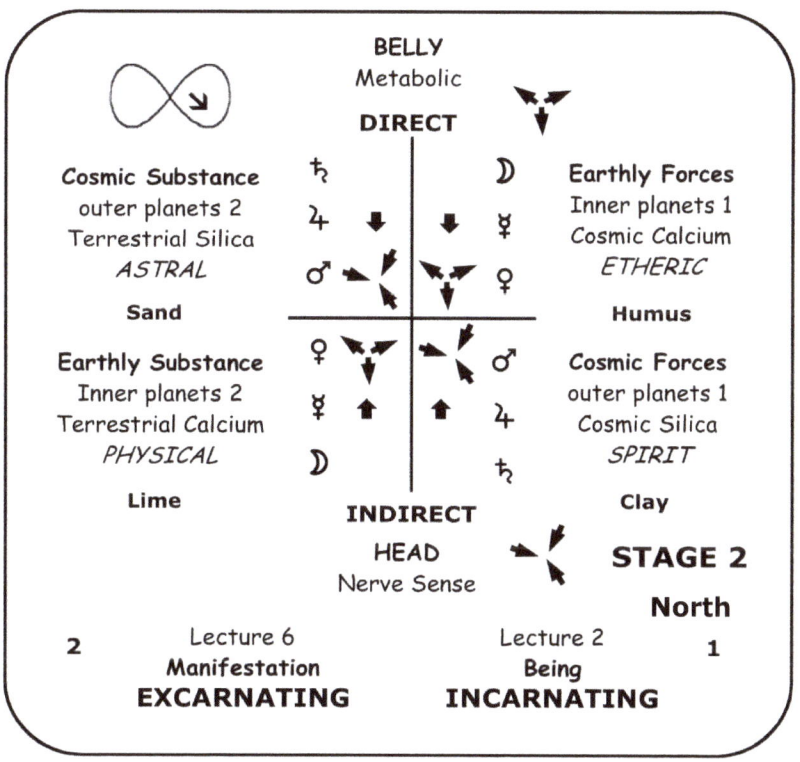

the gyroscope of opposites working together.

We are given half the planetary order in lecture 2 where RS is describing the Incarnating Force side of the story and the other half in lecture 6, where he tells us of the Excarnating Manifesting side of the story. Lecture 6 where he talks of the inner planet activity comes from the Earth Mother and the outer planets come from above as Forces of Sky Father, makes sense. However Lecture 2 trips us up, when RS says the 'male' Outer planets are most active as Forces, when they come from the Earth and the 'female' inner planets are most active as Forces when they are working from above downwards. Here we find a lemniscate twist, however we first need to appreciate that the Outer and Inner planets both form a circular process in nature. The inward moving outer planets come to Earth and are then reflected back outwards, as do the Inner Planets. They both form a continuous cycle through the Earth, Plants and the Atmosphere. The whole plant is not limited to its physical form. The Physical and Etheric parts of the plant are what we see, however its Astral and Spirit activities remain outside it. The Astral as 'cloud' around it , while its Spirit accumulates below it in

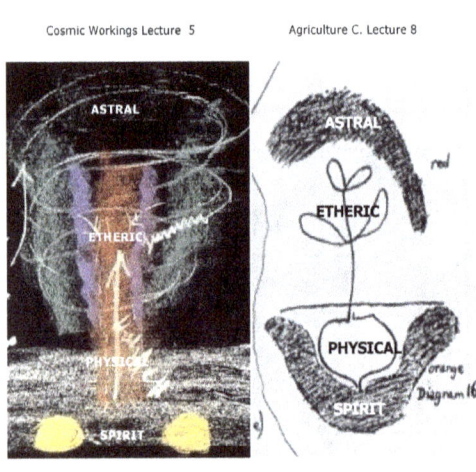

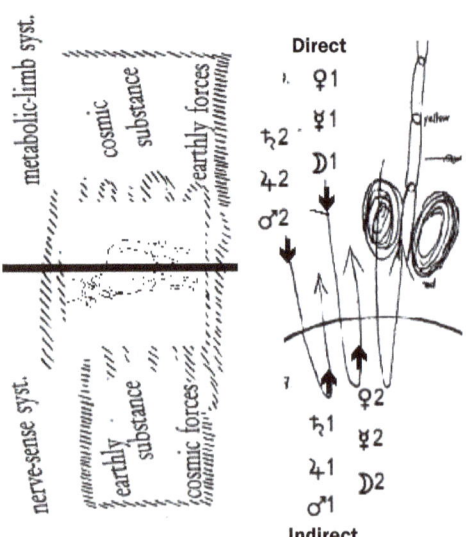

the Earth. (See 3). We are given these two images to represent this.

The key development in the Agriculture Course is we are given the internal organisations of the physical bodies of the kingdoms of nature, which form into the 3 physical organisations. Things are polarized.

RS clarifies a primary polarity for the plant and soil, that the four bodies activities organise. The activity of the soil and roots of the plant are similar to the Head of the Human, with a predominately contractive quality. While the flowers and fruit of the plant, being the reproductive parts of the plant, are reflected in the Human

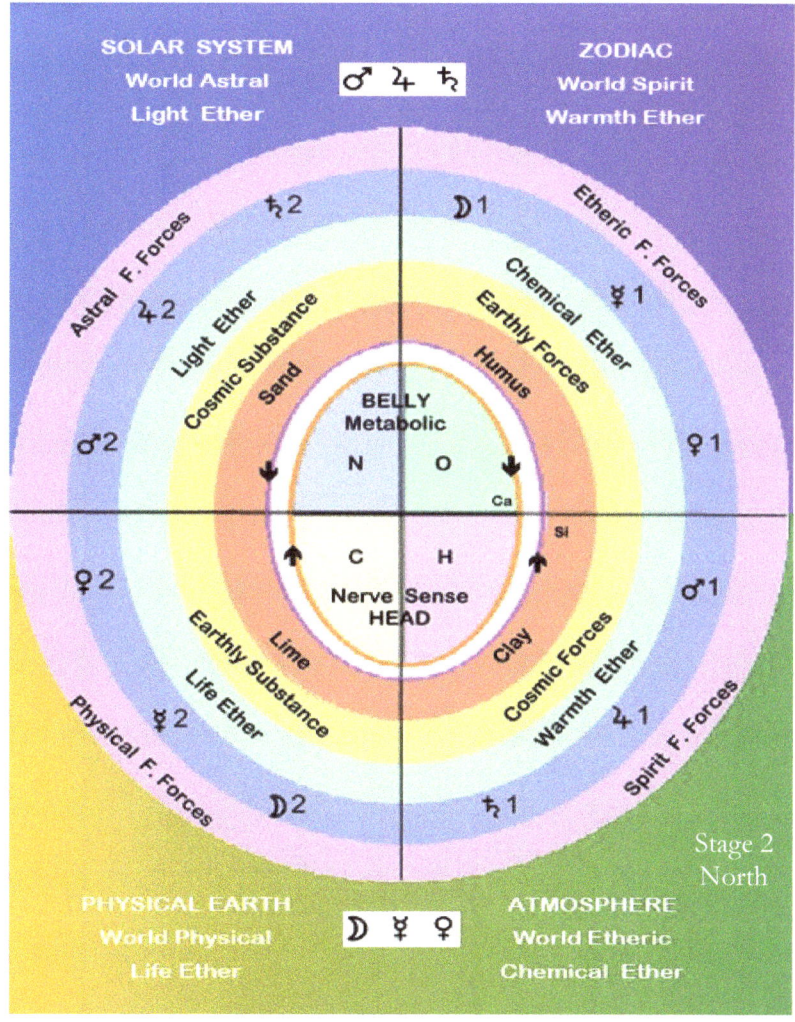

Metabolism, as a expansive influence

We can then see the metabolic system as the Excarnating outer planets 2 working with the Incarnating Inner planets 1, while in the Nerve Sense system, the Incarnating Outer planets 1 are working with the Excarnating Inner planets 2,

RS attempts to simplify this story by moving from the planetary sphere (ring 5) to the Physical body sphere (Ring 3) where he talks of the Incarnating Outer planet activity as Cosmic Forces, the Excarnating Inner Planets activity as Earthly Substance, and the Excarnating Outer planets as Cosmic Substance and the Incarnating Inner planets as Earthly Forces.

In this discussion RS speaks of Direct and Indirect planets, in three different contexts. For the version that most interests us in this discussion he provides this image on the right image from lecture 6 of these activities organization. The left image is from Lecture 8, while . An a enlarged version of this information is below.

It is at this second stage of the creation process, when things begin to move and polarise, that BL and RS begin to talk about Direct and Indirect Planets. These are a Stage 2 and Stage 3 phenomena.

One difficulty of this topic is that RS described three different versions of what constitutes Direct and Indirect planetary activities. I go into the details of these three approaches in the my essay on the 'Layers of the Story'. Each of these three organisations appears to have their specific place of application. The one we are discussing from lecture 6 appears to be most related to the internal growth processes of living beings. In this version, we are focusing upon the Direct forces coming from Above, they are absorbed by the Earth, and then reflected back outwards, as Indirect forces, in the same place. These are firstly 'World' processes, occurring outside in nature, and secondly they work inside lifeforms to influence the way internal activities occur within the 4 kingdoms of nature.

These activities sound very similar to the Incarnating and Excarnating Streams, but NO, these Direct and Indirect activities are only talked of in regards to the Stage 2 & 3 organisations. (See 2 pg. 19).

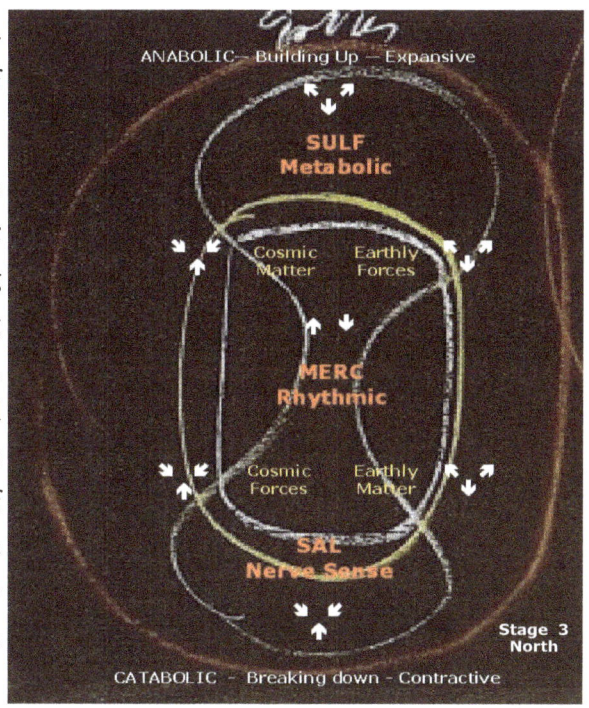

When we put lectures 2 and 6 together, we can see RS has the Direct Forces coming from above, and having two sides. There are forces coming from the Outer planets 2, (those beyond the Sun) which have a inward moving contractive Catabolic action, and called Cosmic Substance. While the Inner planets 1, (those between the Sun and the Earth) are referred to as Earthly Force activities, with their expansive Anabolic quality. Add these, to the two types of Indirect planets, and we have a fourfold form. When they work within the Physical, the Direct Inner 1 are called the Earthly Forces, The Outer Planets 2 are Cosmic Substance, working from Above. The Indirect Inner 2 are Earthly Substance and Outer Planets 1 are the Cosmic Forces, working from Below. Thus we have a expansive and a contractive influence in both the Inward Direct and the Outward Indirect activities. We can also say there is one Incarnating and one Excarnating process in each of the Direct and Indirect streams of activity.

So while we can find a dominant expansive process in the Metabolism, we can also find a contractive growth process, working with it. Similarly within the Nerve Sense system, it has a dominant contractive influence, supporting the contractive Spirit based Cosmic forces, with a secondary expansive influence, coming from the Physical Earthly Substance processes. This expansive influence has 'migrated' from the metabolism, into the head, through evolution, and still works upwards to support the 'nourishment' of the brain substance. Too much of this and we have migraines.

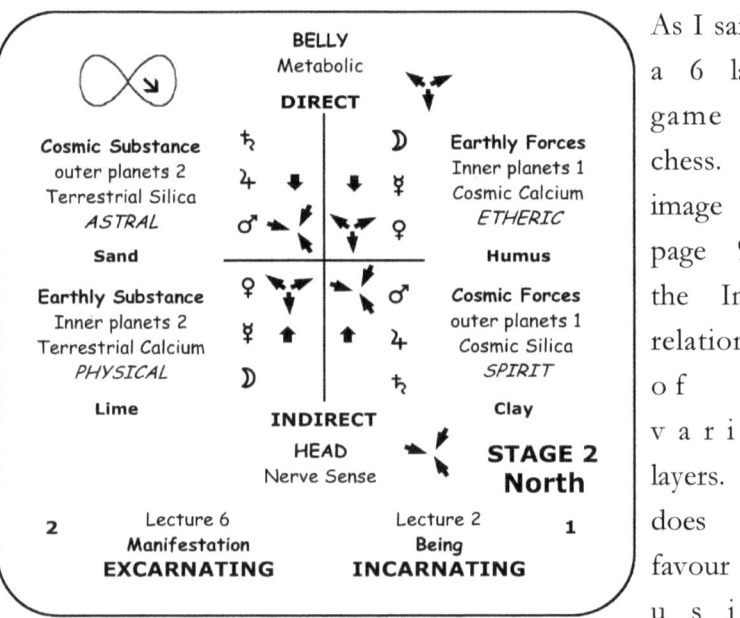

As I said it is a 6 layered game of chess. The image on page 93 is the Internal relationships of the various layers. RS does us a favour by using different names for the activities in the various layers however he only tells us how they all fit together in a couple of places.

The earlier Rings picture (pg. 86) is of the External World, while this image (pg. 93) is of the 6 Internal layers of organization. This is the second Soul World organization standing behind the 3rd World of Manifestation.

The Third World Enfolded Manifestation.

My finding of this organisation occurred over several years, and was clarified from a few lines of observation coming together.

From stage 2 we have the Physical Formative Forces (PFF) organized as the Cosmic Substance (outers 2) and Earthly Forces (inners 1) in the Metabolism, and the Cosmic Forces (outers 1) and Earthly Substance (inners 2), ala page 96

In the Third Word another twist occurs. This is the movement of the lemniscate of the Soul World enfolding upon the Spirit World to create the External and Internal Physical divide. Creation becomes Physical. This is the energetic dimension of Chemistry, the science of physical matter.

In the planetary picture we have been following this enfoldment twist is seen as a twist of the bottom part in the nerve sense system. The Earthly Substance flips over to the bottom right of the diagram while the Cosmic Forces move to the bottom left of the picture. I have found this organization in a few places within manifestation.

The first hint of this order came when I summarised all of RS medical lectures into one diagram. I called this the Glenological Rosetta Stone, due to all the different reference groups, I was able to place together. Bottom pg. 98.

The first thing to note from this diagram is the placement of the PFF. They are not in the Incarnating Excarnating columns found in the earlier two stages. Pg 88 and 91. They are still in

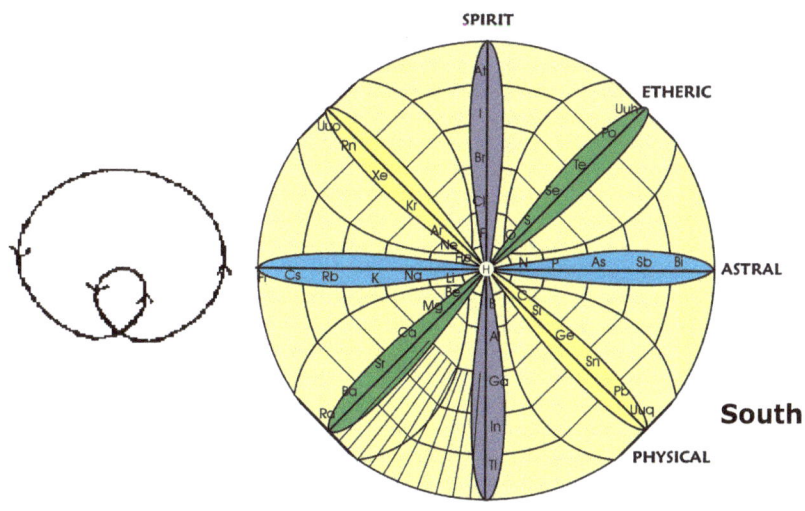

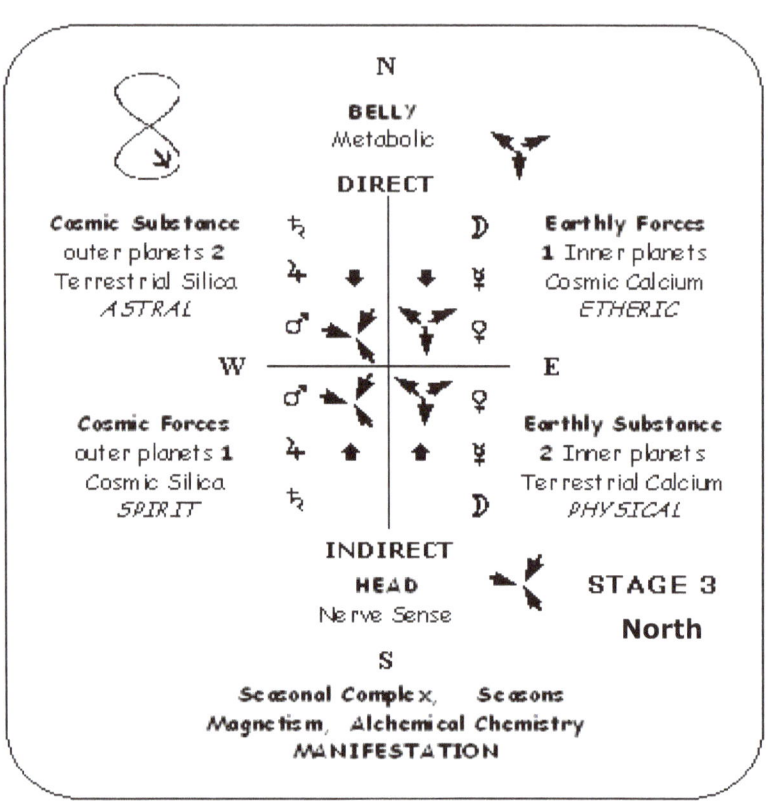

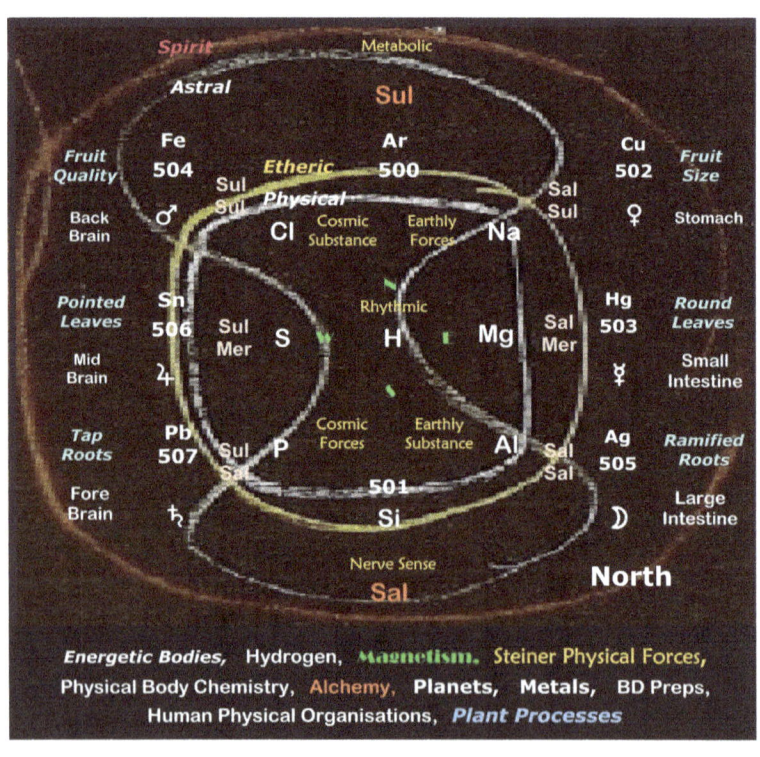

their Direct and Indirect placement.

Another observation is the chemical elements in the next ring out from the PFF. These elements are how they sit on the Physical ring in the Circular Periodic Table. This is a real earthing basis for RS's observations, as it provides a functional reference for the whole of Chemistry.

A third observation is the next ring out and the way the alchemical elements (Sal Sal) are placed in this diagram. These have come from an observation of the picture of the 'Metallic States' of the chemical elements. This image in itself is a 'development' of the Gyroscopic Periodic table. Following the question — **How can we look deeper at the chemical elements activity on physical processes**? Came the answer, orientate the gyroscopic PT on the **Internal Physical Arm**, rather than the World Physical Arm. This provides very fruitful images of the biochemical interactions of the elements, using all of RS's medical insights.

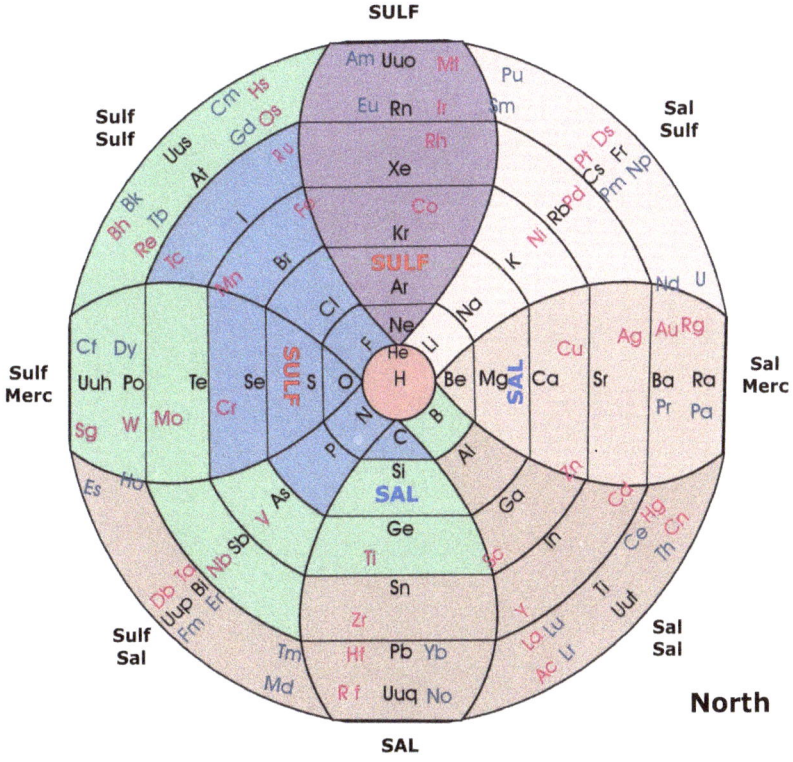

Here we see the Earth Metal elements in brown, the Non metals in green, the gases in blue and the noble gases in purple. We can observe there is a polarity between the bottom metallic elements and the top noble gases, as well as another polarity between the metals on the right and gaseous elements on the left side of the diagram. This provides a framework to look at the 'alchemical' nature of the elements, within a 3x3 form, and much more besides.

Next in my observation of the organisation of the Physical Formative Forces, I came to the stories told in the 'Sap and Elemental Stories' given by RS in October and November of 1923.

These are two very rich lectures that tell the same story of plant growth from two different angles, through the seasons. Reading these I realised my previous PFF diagrams are all static. They tell of the processes and their interactions, but not of how and when

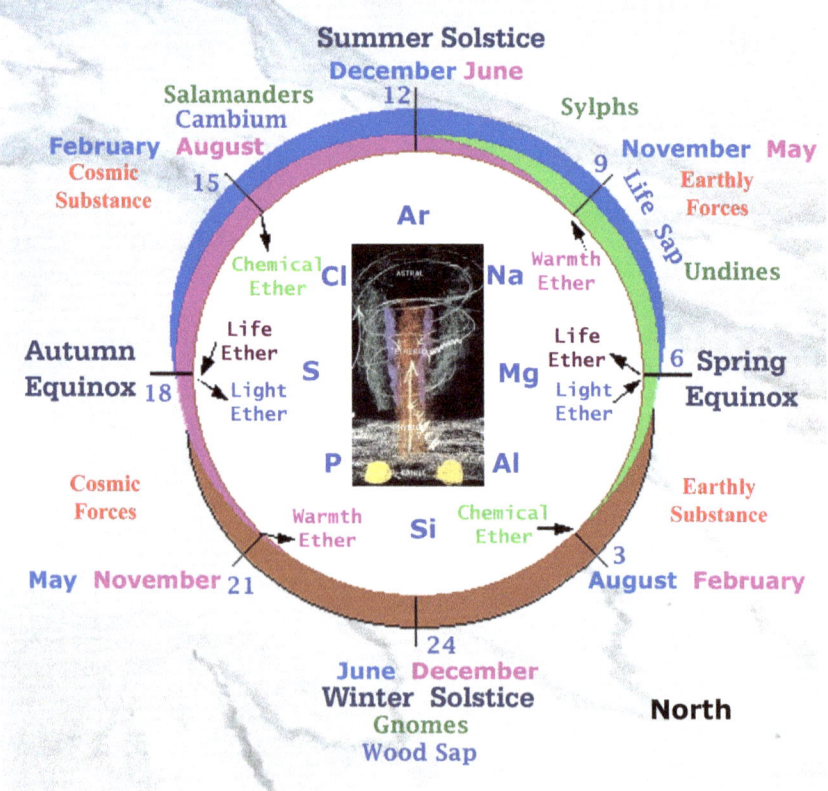

they are most active in the Seasons. These two lectures do that. They are the stories of the PFF in motion, and nature moves through the PFF activities in the same order found in Chemistry and the Glenological Rosetta Stone.

Stage 1 and the Spirit World is Lievegoed' story, Stage 2 and the Soul World is the Agriculture Course and Stage 3 and the Manifest World are talked of in the 1923 lectures. They each tell a different part of the one story.

There is a pathway through this maze.

RS tells us we can control the Physical Formative Forces using Clay for the Cosmic Forces, Sand for the Cosmic Substance, Humus for the Earthly Forces and the Cations for the Earthly Substance. **These are levers we can control the PFF layer of nature with.** However I find it useful to know where I am, and the rules that apply for where I am in this game of 12 dimensional

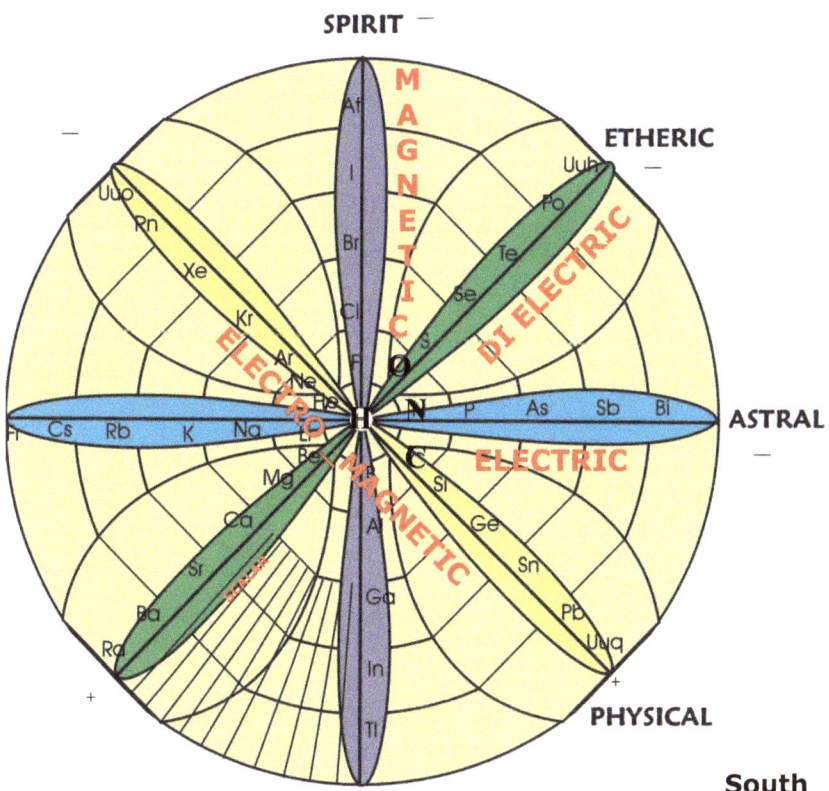

'chess', RS provided us with 6 External Layers and 6 Internal Layers.

For another application of this organization we can look some more at Chemistry, as this is the dimension it manifests. I had looked into the whole Circular Periodic Table, with its 120 elements when there came the question, from Dave Robison of Oregon, as to how RS's comments in lecture 3 about H, N, O, C being the carriers of the energetic activities into manifestation, fit this model. These are the 4 base elements for all carbohydrates and proteins.

The answer to this question came by looking at where these elements are placed on the circular periodic table. Each element sits at the bottom of one arm, whose positive and negative companions of the same valance, and form one axis. So this provides four axis of a positive cation and negative anion dynamic, within manifestation.

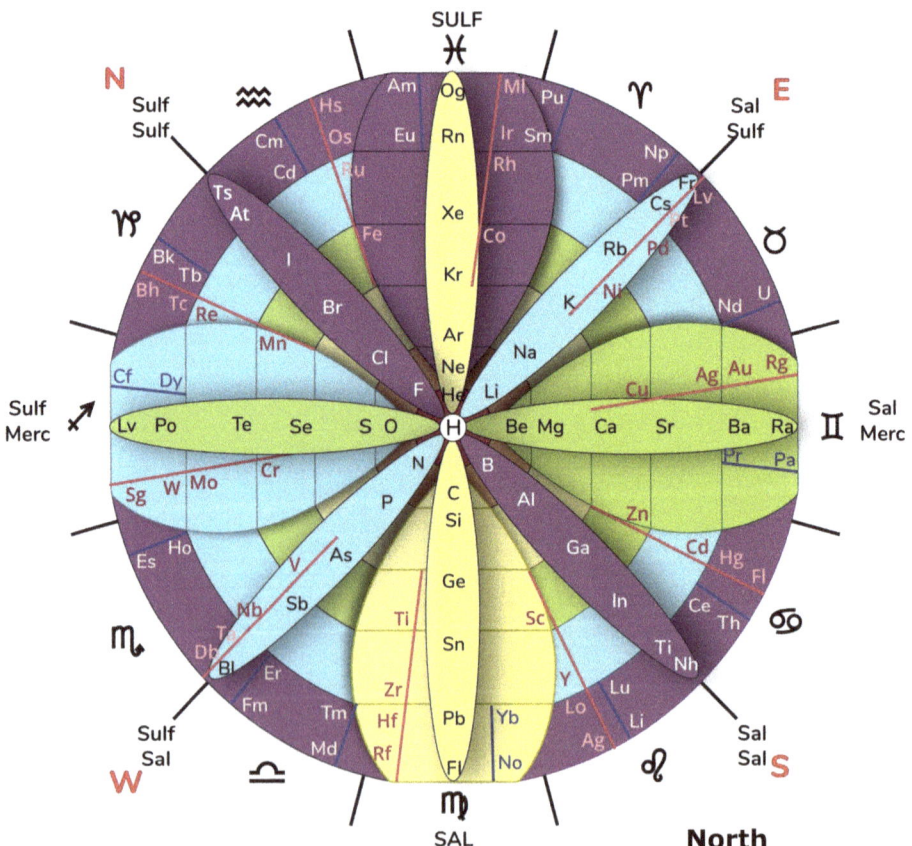

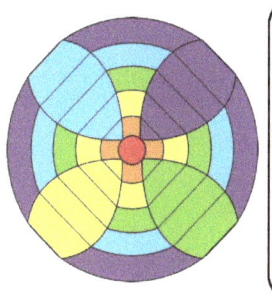
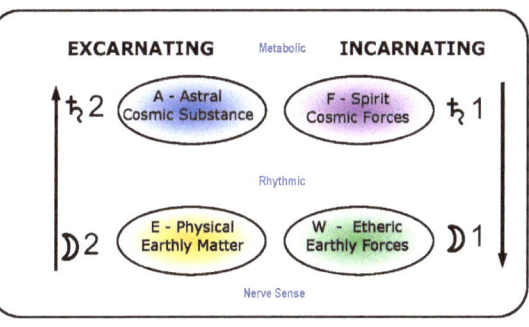
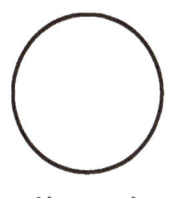

Lievegoed
Stage 1
ARCHETYPE
SPIRIT

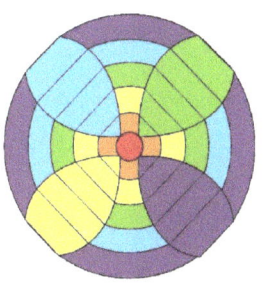
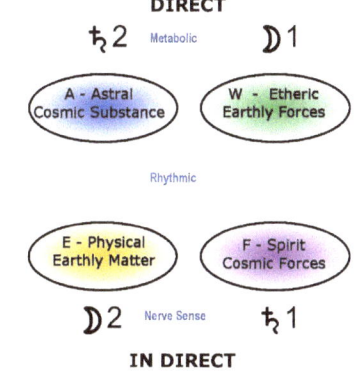
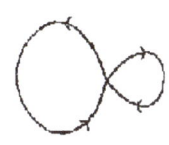

Lecture 8 Agric. Course
Stage 2
BEHIND MANIFEST
SOUL

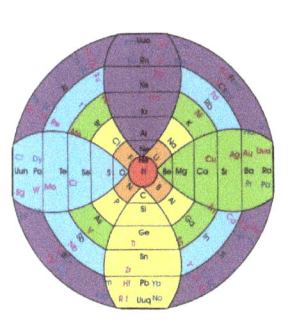
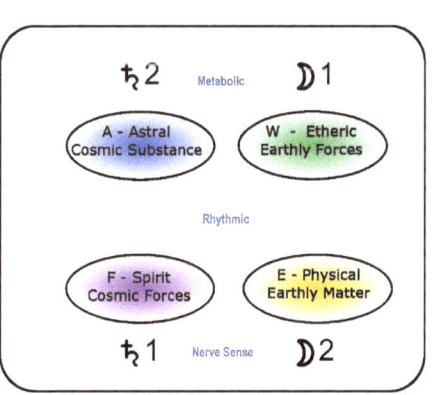
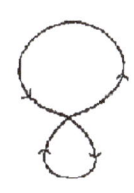

Seasonal Cycle
G. Rosetta Stone
Int. Phy PT
Stage 3
MANIFEST

North

A further observation along this line came when considering the physical realities of Magnetism, Electricity, Electro-Magnetism and Di-Electrics. Following the physical laws of their relationships, as described in their literature, these four Force phenomena of every spinning gyroscope appear to align with the 4 base chemical elements.

I call these axis the Petals. They represent the third Manifest World layer of my diagrams.

The Rings are the Spirit World, the Arms are the Astral Soul World and the Petals are the Manifest World

References

1) Theosophy   R Steiner
2) The Working of the Planets and the Life Processes in Man and Earth   Dr Lievegoed
3) Energetic Plant Growth - www.garudabd.org

The Spirit in Biodynamic Agriculture
GA Italics v5 3.25

The Spirit in Biodynamics is a very rare beast. It is usually heard described as being the human in charge of the farm enterprise, the one who brings direction to the farm. We almost never hear it being a growth influencing force, even when we can see that wild nature has no need for Humans. We hear the Etheric and Astral mentioned in somewhat vague terms, as growth influencers, but where is the Spirit. RS told us everything has the four energetic activities functioning at all times, but somehow plants have lost their Spirit function in present day BD.

When using the four energetic activities as the basis of BD actions, we will come to see the Spirit or Star Forces to correctly address it, is the primary 'architectural' source activity from which all other things arise. It has to be considered as the central driving impulse of plant manifestation. The Etheric and Astral are responding to the Spirits direction. The Spirit manifests in many different ways within our BD worldview.

It is unfortunate we use the term Spirit, as this word has so many connotations for so many people. Many folks give it a religious reference, pertaining to a creator 'father' God, or a more generalized universal presence, or as something that manifests only as a self consciousness within humans. Dr Steiner made an attempt to clarify this self consciousness, from World Spirit by calling it the incarnated Spirit , Ego. However this too has its difficulties given the common reference of ego, to mean 'self centered', and not in a good way.

For a energetic Biodynamics, Spirit is a very real and functional terminology of one of the main levers we use to influence life processes. Most everyone will agree that Spirit is the primary creative force, however I appreciate it as a 'physical' force sourced from the real Stars, and that the constant beam of EM force coming from any particular star, will be received by the Earth, before being reflected back into life, as the archetypal resonance, standing as the 'central organising pillar' of one particular species.

We have three different relationships to the Stars to consider. Firstly, our primary relationship with the Stars, is the one we have with our Sun. The Earth exists in intimate relationship to the Sun, as a manifest compost heap within the Sun's EM onion sphere. The Sun is our parent, the 'being' from which we came, literally, and our life depends upon it. This is called the **World Spirit**. The second relationship with Spirit is beyond our Sun. The Sun is only one of billions of Stars that belong to our Galaxy, which is our ultimate localised organised 'being'. The Sun is its child. Beyond the Galaxy, there are many other galaxies, and thus many many other Stars, all beaming their individual resonance at us, constantly. All of these forces, our Galaxy and all others, are collectively called the **Cosmic Spirit**, but our Galaxy takes central emphasis for us at this level, as our locally organised Star collective. I find it useful to remember the saying, 'there are as many stars in the sky as grains of sand in the Sahara desert', when contemplating the Cosmic Spirit, and how each star can manifest on Earth as a individual species, as suggested by Paracelsus. There are enough stars out there even for every human to have their own star.

Our third relationship with the Spirit, is when the Star forces are reflected back from the Earth, and we see life processes. Here we find the 'internalized' Spirit, holding the blueprint of a Being, and via the processes of evolutionary developments through the species of nature, the **Internalised Spirit** has come to be experienced by us, as our eternal conscious self, which stands behind our comprehension and objective rationalization of the world around us. This is purported to be a purely human experience, however once blood begins to appear in a species, we see the fingerprints of the Spirit internalising. All kingdoms other than the Human have a Spirit activity however it acts onto them from outside of their physical body, as a collective phenomena.

The rest of nature supposedly does not experience Spirit as self consciousness. For most of creation, the Spirit is the 'holder of the plan', with a central organising, and directing role to play out through its more physical processes. If the Spirit 'goes on holiday', an array of illnesses arise in all kingdoms of nature. In Humans, we see such illnesses as anxiety through to cancer, diabetes and dementia. Indeed most modern illnesses are due to a poorly incarnated Spirit. In plants there will be no upright growth or seed formation.

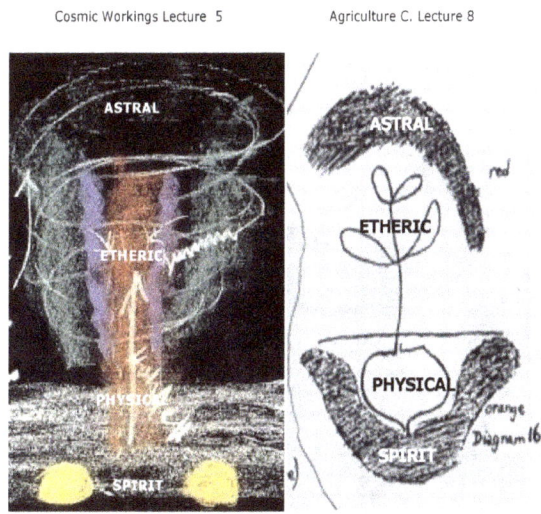

Dr Steiner provides this overview.

"Let us draw the plant in its entirety (Diagram 16). Down here you have the root; up there the unfolding leaves and blossoms. And as above, in the leaves and blossoms, the astral element (red) is acquired from contact with the air, so the ego-potentiality (orange) develops below in the root through contact with the manure. The farm is truly an organism. The astral element is developed above, and the presence of orchard and forest assists in collecting it. If animals feed in the right way on the things that grow above the earth, then they will develop the right *internal Spirit* / ego-potentiality in the manure. If they produce, this ego- potentiality, *it will connect with the plants ego potential and Cosmic Forces, already within the Earth (see gold balls in another RS image) and* work on the plant from the root, causing it to grow upwards from the root in the right way according to the forces or gravity. It is a wonderful interplay. (31)

The plant-world develops in such a way that it represents only physical corporeality, etheric corporeality; that is, in the actual plants themselves. But when we come to the astral element of the plant-world, we must imagine this astral element of the plant-world as an astral atmosphere which encompasses the earth. The plants themselves have no astral bodies, but the earth is enveloped in an astral atmosphere, and this astrality plays an important part, for instance, in the process of the unfolding of blossom and fruit. The terrestrial plant-world as a whole, therefore, has one uniform, common astral body which nowhere interpenetrates the plant itself, except at most in a very slight degree when *pollination* begins in the blossom. Generally speaking, it floats cloud-like over the vegetation and stimulates blossom and fruit formation. (32)

The plant egos / *Spirit* dwell in the very center of the earth, whereas the animal group souls , *and plants astrality,* circle round the earth like trade winds. All these plant egos at the center point of the earth are mutually interpenetrating beings, for in the spiritual world a law of penetrability prevails and all beings pass through one another. We see the animal group souls moving over the earth like trade winds, and how in their wisdom they carry out what appears to be done by the animals. Studying the plant we see that its head — the root — is directed towards the center of the earth where its group ego is to be found. The earth itself is the outward expression of soul and spirit beings. From the spiritual point of view the plants seem like the nails of our fingers. The plants belong to the earth, and when we look at them singly we do not see a complete entity, for the single plant is just one among the whole number of beings constituting a group ego. In this way we can enter into what the plants themselves feel. The part of the plant that springs up out of the earth, what from within the earth strives up to the surface, is of a different nature from what is growing under the earth. There is a difference between the cutting off of blossoms, stalk, leaves, and the tearing up of a root. The former gives the plant soul a feeling of well-being, of pleasure, just as it gives pleasure to a cow, for example, when the calf sucks milk from her udder. There is actual similarity between the milk of animals, and that part of a plant which pushes its way out of the earth. When in late summer we go through fields where corn is being cut, where the blade is passing through the corn stems, then the whole fields breathe out a feeling of bliss. It is an intensely significant moment when we not only watch the reaping with our physical eyes, but perceive the feeling of contentment sweeping over the earth as the corn falls to the ground. But when the roots of the plants are pulled up, then that is painful for the plant souls. (33)

Curiously enough, the spiritual investigator becomes aware that it is generally impossible to consider the world of plants, this wonderful covering of the earth, as something existing by itself. When confronted with the plant he feels just as he does regarding a finger, which he can consider only as belonging to a complete human organism. The plant world cannot be considered in isolation, because to the view of the spiritual investigator the plant world at once relates itself to the entire planet earth and forms a whole with the earth, just as the finger or piece of bone or the brain forms a whole with our organism. And whoever merely looks at plants by themselves, remaining with the particular, does the same as one who wishes to explain a hand or a piece of human bone by itself. The common nature of plants simply cannot be considered in any other way than as a member of our common planet earth.

An outer circumstance might already suggest to us that, just as every stone has a certain relationship to the earth, so also everything plant-like belongs to it. Just as every stone, every lifeless body, shows its relationship to the earth by being able to fall onto the earth, where it finds a resistance, so every plant shows its relationship to the earth by the direction of its stem, which is always such that it passes through the center of the earth. All stems of plants would cross at the earth's center if we extended them to that point. This means that the earth is able to draw out of its center all those *Cosmic* force radiations that allow the plants to arise.

If we now study grain-producing plants, we discover remarkable little organs present in all these plants. Small structures in the starch cells are discovered. These cells are constructed in quite a remarkable way, so that within them there is something like a loose kernel. These structures have the unique property that the cell wall remains insensitive to the kernel at only one spot. If the kernel slips to another spot, it touches the cell wall, leading the plant to return to its earlier position. Such starch cells are found in all plants whose main orientation is toward the center of the earth, so that the plant has an organ within that always makes it possible for it to direct itself in its main orientation toward the center of the earth. This discovery, made during the nineteenth century by various scientists, is certainly wonderful, and it is most remarkable if it is simply presented as it is. Natural science *also* shows us that mistletoe does not have those curious starch cells that orient the plant toward the center of the earth.

But now let us turn to something else. If the leaf of a plant is studied, it is discovered that the outer surface is actually always a composite of many small, lens-like structures, similar to the lens in our eye. These 'lenses' are arranged in such a way that the light is effective only if it falls onto the surface of the leaf from a very specific direction. If it falls from another direction, the leaf instinctively begins to turn in such a way that the light can fall into the center of the lens, because when it falls to the side it works in another way. Thus there are organs for light on the surface of the leaves of plants. These light organs, which actually can be compared with a kind of eye, are spread out over the plants, but the plant does not see by means of them; rather the sun being looks through them to the earth being. These light organs bring it about that the leaves of the plant always have the tendency to place themselves perpendicularly to the sunlight.

In this — in the way the plant surrenders itself to the sun's activity in spring and summertime — we have the plant's second main orientation. The first orientation is that of the stem, through which the plants reveal themselves as belonging to the earth's self- consciousness; the second orientation is the one through which the plants express the earth's surrender to the activity of the sun beings. You will find little by little how the plant covering of our earth is the sense organ through which earth spirit and sun spirit behold each other. The carbohydrates can arise only if the sun spirit and the earth spirit kiss through the plant being." (30)

Spirit in the Environment

Dr Steiner has provided us with several stories of how the Spirit works within Life. The cornerstone image of his worldview, 'As Above , So Below' suggests whatever occurs on Earth is but a reflection of what is going on above.

We first meet the Spirit, in the Agriculture Course in the first lecture. However it is not very obvious. In this lecture he talks of the two basic streams of activity, the Cosmic and Earthly streams. The Cosmic stream is all that comes from above, while the Earthly stream is what comes back outwards from the Earth. He comments that the Cosmic stream starts with the Stars and it uses the outer planets, warmth and light, and silica as its physical carriers. What is not made totally clear in lecture 1 is that the Cosmic stream includes both the Star based Spirit forces, and the Planet based Astral forces. Wherever the cosmic stream goes throughout RS stories, both of these are present, however depending on the place this stream is working, one or other of these partners will dominate. From lecture 2 onwards his stories talk of Cosmic Forces, where the Spirit is dominant, and Cosmic Substance, where the Astral is the dominant partner.

The second lecture tells how the Cosmic and Earthly streams work into and out of the Earth, and through plants, however for the sake of comprehension, it is best to talk of this a little later. A better place to start is the third lecture, as it provides a very useful and practical indication of the four primary activities and their physical carriers. Sadly, his description is not as clear there, as when he tells a similar story in his medical lectures.

In the third lecture, he outlines one of his greatest gifts by showing how the elements of protein act as the physical carriers, of the four astronomical energetic spheres, into life. We only need to look at the reality of a star to know that Hydrogen, the lightest, and first element of chemistry, is the base fuel by which a star runs. A star's very real EM forces, arise from the nuclear generator at its center, fueled by the combustion of Hydrogen and its first 'child' Helium. So wherever we find Hydrogen in nature, we have the physical tracks of the star forces, we call Spirit. It follows that nitrogen, found around the planets, eg our atmosphere is 80% N, is the physical carrier of the planetary activity, we call Astrality. Oxygen, found in a free state in our atmosphere (20%), and liberated by plant life processes, is the physical carrier of the life giving Etheric, while Carbon, the physical basis of life forms, is the carrier of the Physical forces. Sulphur – the 16th element of chemistry – acts as an 'oiling agent', allowing these four elements to combine and work together, into biochemistry. To little Sulphur and the bodies become stuck and in Humans we see autism, while too much Sulphur and the bodies can not take hold of each other, and we see hysteria manifesting. In lecture 3 RS did not tell us of Hydrogen's incarnating 'architect' function, he only talked of its excarnating function.

In aerobic life processes, we find that Hydrogen binds firstly with Carbon, to form methane before binding with Oxygen, to form the many carbohydrates. Later nitrogen joins in to form the proteins. Here Hydrogen provides the plan for the Carbon and

Oxygen to use in their building activities. In lecture 3, RS only describes the anaerobic pathway in his description of the elements relationships. (C > N > O) The incarnating role of Hydrogen needs to be considered, as it is the bringing of the plan to the contractor and the workers.

To carry on this building metaphor for a moment, we can see Nitrogen – the planetary element - as the great energizer. Consider, the planets are the only moving part of the game. The Stars – while we know they are moving – the distances are so vast that within our life's experience they do not appear to move, and hence we call them 'the Fixed Stars'. We also know that if water – a carrier of the Etheric - is left to itself it will become stagnant. It needs outside influences to move it so it can stay alive, by activating and adding Oxygen. The Earth, we can see also sits there and slowly decays. So from our experience, the planets are the only moving parts, and their kinetic alteration of the Sun EM body, as they move through their Solar Onion EM Spheres, leads to alterations in both, a) how the Star forces are altered on their journey through the Solar system EM body, and b) how we experience the other planets formative influences, due to their alteration of the Earths magnetic field. It is nitrogen that provides the physical pathway for this planetary Astrality to move and activate all other parts of the life processes we deal with in nature. However, Astral activity is chaotic if left to itself. It will follow the EM wanderings of the 9 planets, with no limits. In Buddhism the astrality is characterised as a beast that needs to be identified, tamed and then used. In lifeforms, it needs a plan to follow, and this comes from the Star 'architect' ,who provides the direction to the 'head contractor', the Planets, who then use the Atmospheric Etheric 'workers' to do what is needed to the Earthly materials they have to build with.

While the Astrality is 'the energiser', its energy is more of the nature of enthusiasm, and a hyper activity often seen associated with the many nitrogen alkaloid drugs. We need real food to stay healthy, and this comes via the Etheric and oxygen. Nitrogen readily attaches to Oxygen so it can stabilize as NO2, similarly the Astrality uses up the Etheric /oxygen during its 'enthusiasms'. In our rhythmic system the lung processes are maintained by the astral related Potassium, which needs to be continually balanced by the circulatory systems Magnesium based Etheric activity. If Potassium levels in the blood become to high, the Astral dominants and heart palpitations and heart attacks, are the result. These two poles need to continually work together, to keep each other from excessive expressions.

We find this basic arrangement of the bodies outlined above in all kingdoms of nature. Even though the lower kingdoms do not have an incarnated Spirit, this 'Star' principle still plays onto those kingdoms as a collective influence, directly from the stars. It is experienced as we experience warmth. It is all around us and we are encased in it, and so we have access to it. This influence starts with the species type, and can be seen in the collective organisation of the 'flock, mob or hive'. Some people call this external organisation a Deva. Beyond this, the tracks of the Spirit can be traced through its working into the other energetic spheres, that are incarnated. Given everything is electro magnetic, all the energetic activities have to work into each other. The Spirit works into the Astral, Etheric and Physical spheres, while the Astral works into the Etheric, and Physical. All the lower activities also push back outwards into the spheres above. All these interactions leave tracks we can follow. Once we understand them, and their relationship to chemistry, we can use them as part of our energetic bag of tricks.

Spirit in the Astral Formative Forces

The incarnating and excarnating journey of the Spirit working with the other activities can be seen in more detail when seen from the perspective of the planets, as told by Dr Lievegoed.

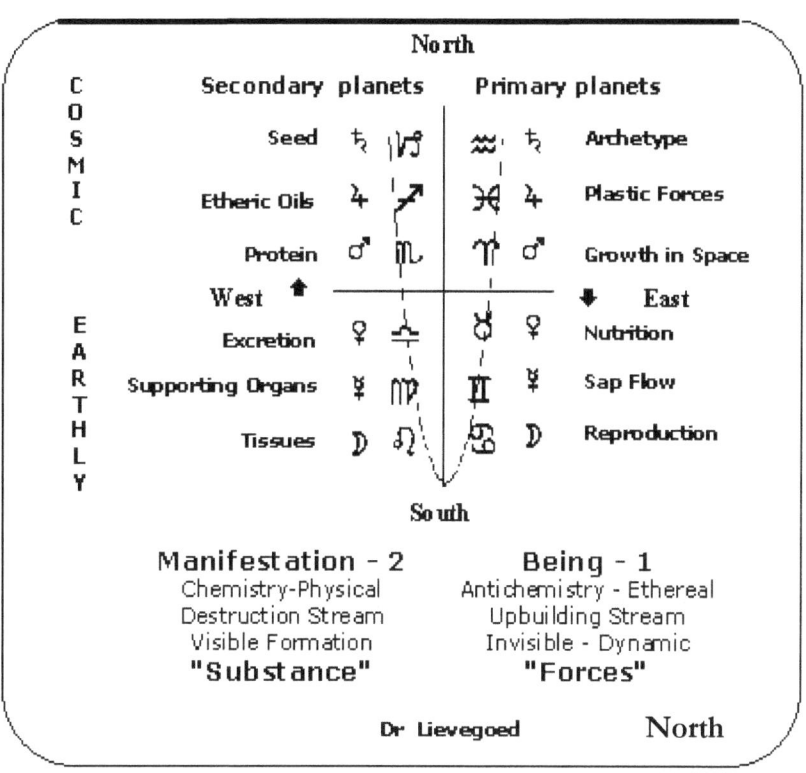

The Cosmic and World Spirit resonance's (the plan) enter the planetary sphere and are taken hold of by Saturn 1. He becomes their servant, as the 'astral representative' of the architect on the board of the contractors company.

Jupiter 1 is the engineering head, who brings the Spirit and the Astrality together by making suggestions as to how best to adapt the plan to the requirements of the site. Mars 1 is the job foreman, who needs to organise the schedules for the actual work to begin. At Venus 1 we see the Etheric accumulating the required resources and materials. Mercury 1 brings all the workers together, while the Moon 1 makes sure the physical site is ready. This whole phase is the preparation for something to happen. This is the 'coming into Being' phase 1 part of the cycle. Work, birth or germination is about to begin.

The outward moving 'manifestation' phase starts with the growth processes of Moon 1 which wants to just grow endless cells, being directed by Saturn 1's plan, to form themselves in tissues,

(Moon 2), so more organised growth can begin to form. Mercury 2 provides the fluid movements needed for these tissues to be sustained, by facilitating the Etheric activity, while Venus 2 provides the nourishment for growth to be sustained. This included the flow of minerals and water upwards but also the products of photosynthesis back to the roots. The processing of waste products is also done here. Venus 2 brings the plant to flowering, but it is not until Mars 2 with the introduction of the light filled astrality that we see the formation of proteins as pollen etc. and the actual act of fertilization occurring. Jupiter 2 brings in the atmospheric light and warmth, leading to the formation of fats and other alkaloids, along with fruit quality. It is the Saturn 2 process that sees the contracting forces dominate, and seeds ripen, so the stars plan can take one step further. Behind all of these processes Saturn has stood holding the basic plan in place, from which the other planets could divert a little way. If the etheric activity is carried further into the fruiting process, for example, a certain amount of flavour is exchanged for larger fruit, and more leafy subsequent generations can result. An upright annual plant such as Chamomile Officianalis, can start to exhibit perennial and sprawling characteristics of its wild cousin, but the core plant type remains. (3) Saturn 1 holds the plan of the species all the way through.

Spirit in the Etheric Formative Forces

The layer of our creation below the planets is the magnetic field of the Earth, within which our atmosphere is found. Within Biodynamics this is the realm of the World Etheric and its parts are the Ethers and the Elements. Much has been written and investigated about this sphere, even though it brings the most confusion to the overall BD theology. This is due to some people placing this realm as central to the whole BD story, when it is only one of the six main layers of activity we have to work with. It is not appropriate to go into this topic further here, as I have addressed it elsewhere.

In its journey through the Atmosphere, the Spirit is best seen in the working of the Warmth Ether and the element of Fire. In RS creation story of our Solar System, warmth is the first state from which all other states arise. This period is called the Old Saturn period, due to the primary ball of warmth extending out as far as Saturn. Marti says this is when time originates, and that the warmth ether developed during this period, making things come into existence, and furthers their development and brings them to maturity. The stages of time through life such as the change of teeth and puberty are bought to us through the warmth ether. Van Gelder emphasises 'the impulse-creating warmth of enthusiasm that occurs as the intention that underlies action' and provides directionality to life. (The astral enthusiasm talked of earlier is more of the excited hyper active kind, rather than the spirit intentional energy indicated here)

All of these images as similar to those talked of earlier as qualities of Spirit.

While the Ethers have become the cornerstone reference for a Wachsmuth influenced Goethean observation of nature, - without any reference to the Astral or Physical Formative processes - RS gave no preparations that directly relate to them, and no specific tasks one can take to effect them, other than manipulate the physical elements of warmth, light, moisture, and soil quality, we place around a plant. Which are all standard tasks of gardening. Within the Agriculture Course there is only two clear undisputed references to the Ethers. Their story is told within the lecture series 'Man as Symphony of the Creative Word'. (28) This is probably why the Etherities can not understand the Agriculture Course after 100 years.

Spirit in the Physical Formative Forces
RS italics

While the Ethers are the workers that stand two steps back from

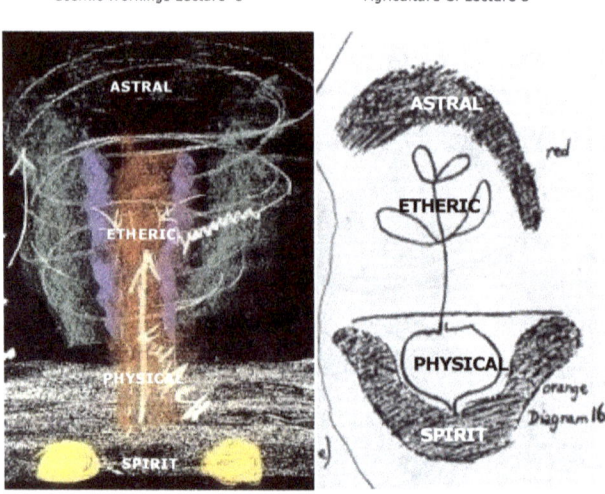

matter, and bring growth energy from the atmosphere to the plant, it is the next layer down in matter, that we find the Physical Formative Forces. These processes stand just behind substance, and are directly involved in the movement of substances. A significant hint of the Spirits role in this sphere is in lecture 8 where the following picture is provided with these accompanying words.

- *"In man the greatest possible quantity of intestinal dung is transformed into cerebral excrement because man bears his ego on the earth. In animals the quantity is less. Hence there remain more forces in the intestinal excrement, of an animal which we can use for manuring. In animal manure, there is therefore more of the potential ego element, since the animal itself does not reach egohood. For this reason animal dung and human dung are completely different. Animal dung still contains ego-potentiality. In manuring a plant, we bring this ego-potentiality into contact with the plant's root. Let us draw the plant in its entirety (Diagram 16). Down here you have the root; up there the unfolding leaves and blossoms. And as above, in the leaves and blossoms, the astral element (red) is acquired from contact with the air, so the ego-potentiality (orange) develops below in the root through contact with the manure."* (more on this later)

Here we are bought to the image of Human manure being of a fundamentally different nature to animal manure, due to the internalised Spirit of the Human, using up all the 'ego potential' left over from the animal. The suggestion follows that if we are to use human manure, for agriculture, we need to put it onto pasture so it can be taken up by plants and then feed to animals, before we use this manure for our own food crops. By doing so the plants concentrate unused Astral and Spirit forces, from the environment. The animal uses the astral forces, in the formation of protein and organ systems, but leaves the Spirit forces unused, as it is not rationalizing its existence, and writing articles such as this one. We can benefit from these leftover forces, in our lively thinking and directed willful actions, when we 'collect' them from plants and animals we eat.

But not all of natures growth is based upon manure. So Spirit forces are not only available from manure, they must be in nature already for the plants to be able to accumulate them. The story of how the Spirit forces accumulate in the soil and root zone in the first place is given to us in the second lecture. After the initial description of the threefold 'agricultural Individuality', there are

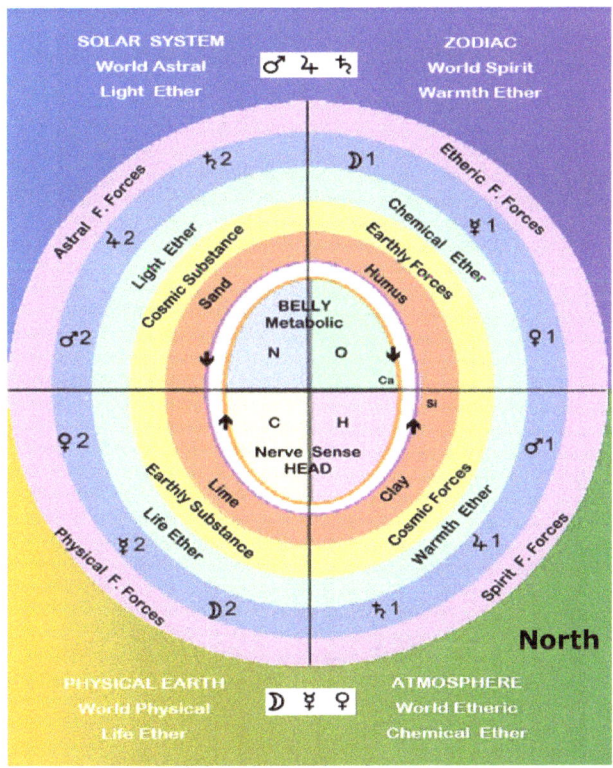

14 pages (of the green edition) describing how the four energetic activities work, as the Physical Formative Forces, into the Earth, through Sand, Clay, Humus and the alkaline Cations. This seasonal based story is enlarged by two other lectures given a week apart in 1923. Each of these lectures use a different language to tell the same story.

Firstly we are told of the Cosmic streams cycle, followed by the Earthly processes journey. In lecture one, the External activities are outlined, and we are left to assume the cosmic activity is **Clay** coming from above. Its source is Spirit and Astral spheres, which work through the outer planets, the elements of light and warmth, and the siliceous substances. So wherever the cosmic stream goes so goes the Astral and Spirit.

In lecture two the story begins with the autumnal light and warmth moving into the Earth with the help of the siliceous sand, within the soil. We need to appreciate that the Cosmic and Earthly processes are each a cycle, moving from one season to the next. RS chooses to start with the autumnal part of the cycle, which corresponds to the ending stages of plant growth and thus the Manifest phase 2 of the planetary story given earlier.

Once we look into the physical Seasons story, we can see a development in the structural order of the Physical Formative Forces, from that given in the Agriculture Course. The seasons story is told in the 'Sap Stories' and Elementals and Ethers stories. I call this change of energy orientation, moving from Stage 2 to Stage 3. The Direct and Indirect planets stay the same, however a further flip of the Indirect group can be observed. The Cosmic Forces and Earthly Substance change places providing us with the flow of these activities that show through the seasons. This order is most strongly indicated as a 'formative' pattern, when we observe the chemical elements 'Metallic and Gaseous States' , and follow the Chemical Elements through RS medical lectures, and onto the Sap Stories. As part of the Alchemical

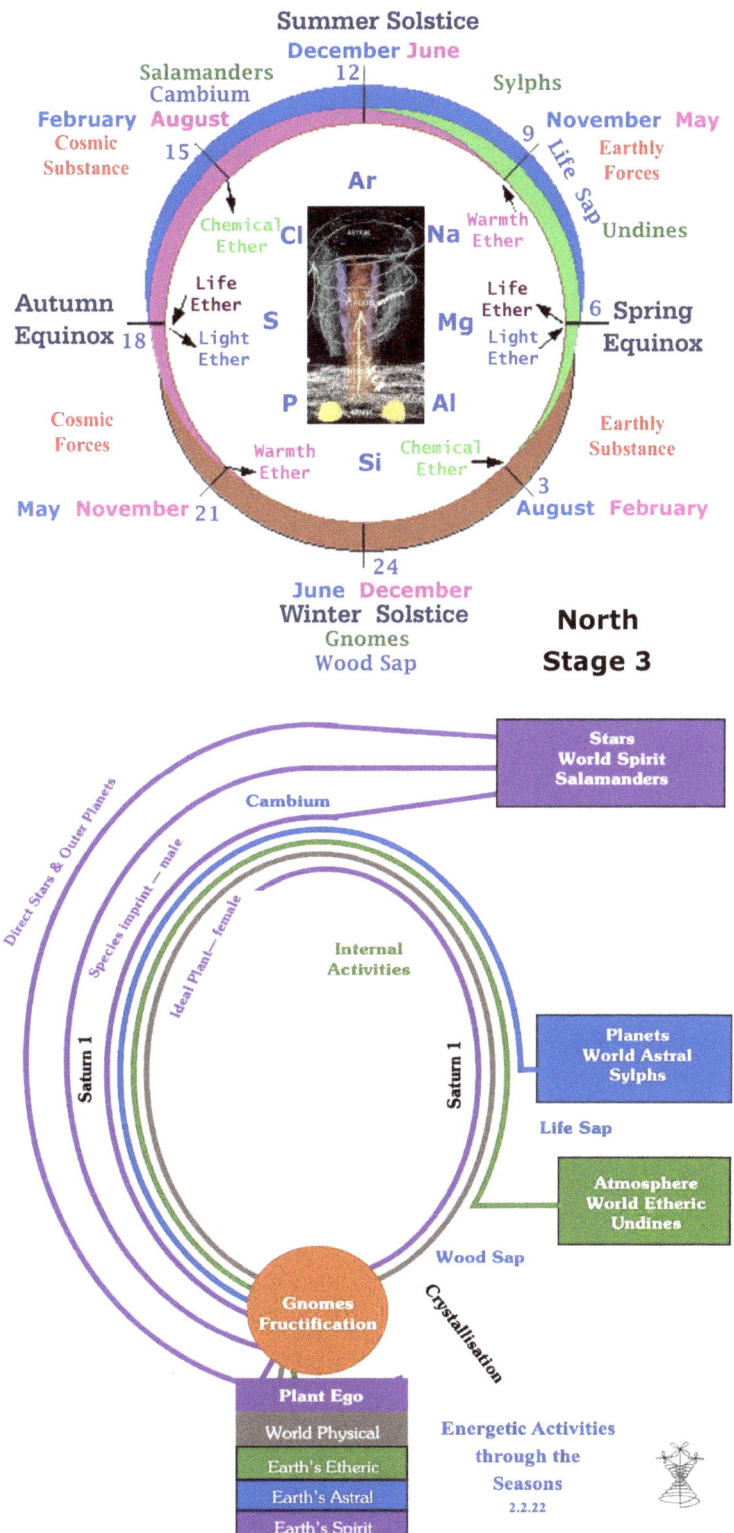

Chemistry story, my 'Rosetta Stone' picture arises, which connects the Physical Formative forces to the Chemical Elements, and to the Seasons. **Stage 3 is** an image of **"What is Manifest"**

The inward part of the cosmic cycle reaches a peak of consolidating activity during the time RS calls 'crystallisation'. This occurs in the weeks after the shortest day, in the northern hemisphere, and appears to be related to the Earth crossing over the Sun's path. This inward stage, sourced from the above ground, metabolic sphere, is firstly called the Cosmic Substance phase of the cycle, within which the Astrality is the dominant partner. In Rudolf Steiner talks of the warmth processes coming from the Stars being drawn into the plant, through the growing season and manifesting, with the other growth processes, as the 'female' cambium processes, that pushes the root exudates into the soil. This 'ideal plant' seeps into the Earth, in Autumn . The gnomes gather up all of this material and energetic residue of this seasons best possible plant. A plant appropriate to its growing period. A Phenotype of its time.

The Star stream, has a second 'male' pathway, where the star forces that carry the new species impulse, enters at the seed chaos phase of the plant, which occurs at pollination. This keeps the rose being a rose. This is the annual reinforcing of the Genotype.

We are told how during the growing season, the Earthly forces want the plant to deny the Spirits plans and spin off in all sorts of directions. This Spirit pathway through the seed, keeps things on track, all the way through to the next seed set, while the Earthly female stream wants to keep it changing throughout the growing season. Saturn 1 keeps the Spirits message on track.

A third pathway sees the Cosmic Spirit forces taken up directly by the Earth and collects 'at the center of the Earth', as the Plant Ego, and from there it can work back outwards. During winter all these streams of activity are gathered up, along with Etheric and Astral Earth forces, by the Gnomes, in the Fructification event, to make the conglomerate 'Wood Sap'.

From the crystallization point the cosmic 'Saturn 1' stream must again work upwards, as part of the Wood Sap, through the plant and all the way through to the new seed formation, for the complete plant cycle to be fulfilled. RS suggests (three times) that if this upward cosmic activity, or Cosmic Forces are not strong enough, we need to add clay to our soil to help it along. RS gives an example of a dominant Earthly substance and a weak Cosmic Force stream, when he talks about fungal problems in lecture 6 of GA 327. Which means the Cosmic forces are being over powered, and why he says the problem is too much vitality. The over active Earthly Substance is not being harnessed by the Cosmic Forces.

A fundamental characteristic of the Cosmic activity, is that it works as a contracting 'anti clockwise' force, while the Earthly processes work through the clockwise expansive spiral. As the cosmic activity comes from the stars, the cosmic stream moves from the periphery to the center, hence its autumnal movement into the earth to crystallization. Wherever we find the cosmic process we find contraction. In the root zone it forms tap roots, as it moves through the leaf zone it causes leaves to serrate and become pointy. In the flowering period it brings the contraction of the anthers and the development of pollen, the deep colours to the flowers, while within the fruit zone its warmth activity brings the roundness to fruit and seeds, but more importantly it contracts, through a dehydration process for the ripening of the seeds.

The contracting processes are played off against the expansive process bought by the Earthly alkaline physical and etheric activities, at every stage of plant growth.

In the below ground 'nerve sense' region, which corresponds to our head, both the activity of the astral and spirit can be found, however in the 'thinking individual', or the tap root of the plant, the Spirit is playing a dominant role. In the metabolic region of lifeforms we will find the astral and spirit are both necessary, however the astrality is the dominant overall player of this pair, there. Hausmann, in his medical writings talks of the solution for shrunken kidney diseases being bought about by the Internal Astrality being needed in the metabolism to stimulate the Internal Etheric into action, and thus produce a dominant catabolic expansive movement, in that region. If the internal astrality is not strong enough - due to an overly active World Astrality over powering it - the internal etheric becomes sluggish and various circulatory problems arise. In this relationship we are shown an image of the interaction of the Cosmic Substance and Earthly Forces outlined in various ways throughout the course. This is also the picture of Equisetum RS asks us to research.

The Spirit needs to be present in the metabolism as well, to bring in the warmth needed to contain inflammations which arise here, and to keep the Astrality under control, while maintaining the processes associated with the liver, gall spleen and pancreas, however overall it plays a secondary role to the Astrality. In the nerve sense system their roles are reversed.

We know a plant will not grow on light alone. Warmth must also be present. So, the light forces above the plant carry the astrality, within which we will find the birds and insects, however we must not forget the role of warmth and the activities this facilitates, such as seed strength and fruit flavour. The larger mammalian predators draw upon the warmth of the atmosphere for their existence.

It is through the Autumn period, that the Cosmic partners change dominance as they move within the soil. The Spirit activity becomes dominant as the 'Cosmic Forces', strengthen through early winter ready for the upward phase of the cycle. This term appears elsewhere in RS literature and refers to the forces that comes from above, (both stars and planets) and within which we live, and which can be detected with our back brain, as 'cosmic imaginations'. This is the constant stream of random images that cross our 'minds eye', which we consider as thoughts, and dreams. It is with the help of Phosphorus that we can bring these cosmic imaginations into our front brain to make them rational objective thoughts or clairvoyant predictions, depending on our skills.

RS use of this term in lecture 2 allows us to follow these cosmic gifts through the plant. It is in these forces (Outer Planets 1 and especially Saturn 1) that the 'will of the species' is being carried, so that the seed that is finally produced will stay true to type – Genotype. It is possible to alter the astral and etheric parts of this journey through the plants life, and bind in other characteristics that do show a divergence from the parent giving rise to the Phenotype. However no matter what changes one makes to the colour of flower or size of fruit etc., an apple tree will only ever produce an apple. This is the Spirit / Cosmic forces keeping things on track.

For healthy seed formation, both parts of the cosmic stream need to meet and intermingle. If this upward stream, fueled by the spirit is not strong enough, and 'runs out of steam' before it reaches ripening, fungal rot attacks will become evident, usually from the time of flowering onwards. They do occur earlier from this same cause, such as 'damping off' disease. This is due to the Earthly Substance processes , from below, dominating. If the warmth and light does not come from above, then whatever is bought from below will not be 'cooked ' sufficiently and alkaloids are not produced, with the seed not contracting enough to retain viability and strength of germination in the following spring. If the Astrality works in too strongly we see powdery mildew and rusts forming, along with all manner of pest attacks.

While I have told the 'cosmic' story here, the whole fungal story is only understood when the Earthly poles journey is added to what is said here.

So back to the manure's ego forces. The 'Plant Ego' is the primary spirit activity sitting below the plant, in the siliceous substances. The plant takes this up, along with Earthly activities in the early Spring. If this is eaten before it gets to accumulate some 'atmospheric spirit' from the warmth and seeding processes, this materialised spirit activity will be given onto the animal, and remain unused to be given back to the Earth in the manure. Having been 'worked on' by plants and animals it is fair to assume that this is a more sophisticated form of Spirit, than that found coming directly into the Earth from the Stars, and intensified through crystallization, but nevertheless it is still a Spirit activity. The difference between these two forms of Spirit, could be imaged as the difference between the octaves of C in music. Same tone, but a different harmonic frequency. Or the difference between the experience of taking homeopathic Silica and drinking a fresh carrot juice.

The Spirit in the mineral sphere. (see p197)

While Silica is a primary element for carrying the Spirit, I have spoken of its role in the Equisetum essay, so the following section deals primarily with the elements of the third 'Cosmic' ring of the Periodic Table. All of the elements in this ring are considered light or warmth bearers, who often burn with a very bright light, often cold, eg Sodium, Magnesium, Aluminum, Phosphorus and Sulphur, or in the case of Silica act as a light enhancer. Chlorine carries this light and warmth process so strongly that it quickly burns anything it touches. This high light phenomena is best understood when we place planetary rulers over the 12 rings of the Periodic Table. Twelve is a twofold division of the 6 primary rings. These elements sit on the fifth ring based on this division. This ring is ruled by the Sun, which is the primary source of World Spirit activity. All these elements are THE light bearers. In the overall scheme of the Periodic Table this is the ring of the 'Cosmic' physical body, and so this is where

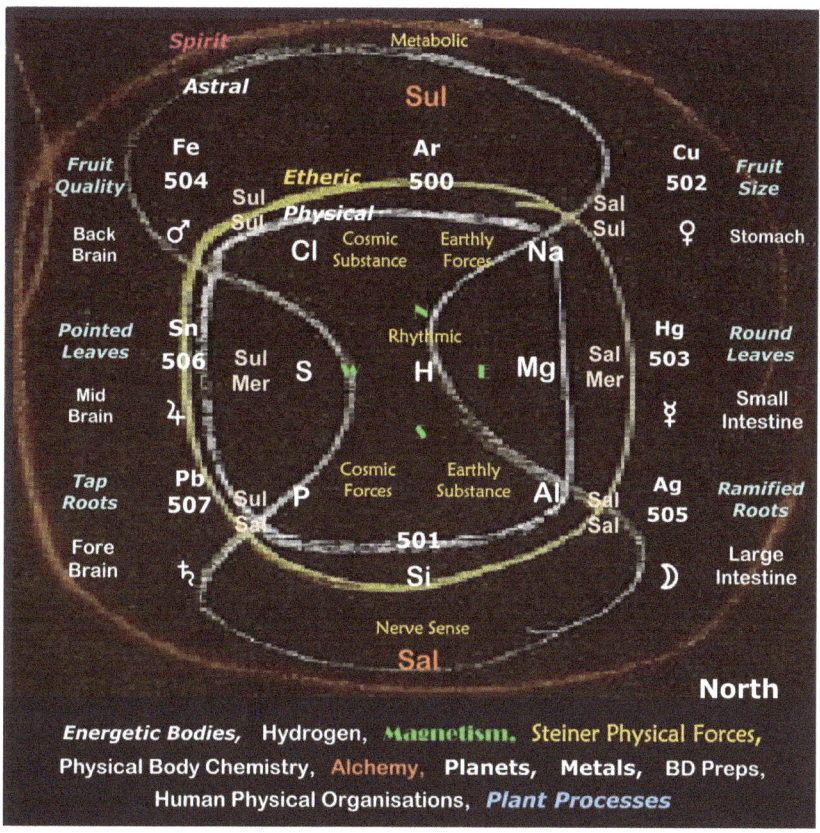

the primary energy of creation is carried into the physical body, thus allowing manifestation to occur. The elements of this ring are all central elements in anchoring the life processes, as shown in my 'Rosetta Stone' image. These are the elements Dr Hauschka uses in his discourse on the elements in the Atmosphere, Hydrosphere and Geosphere organisation.

Clay

While Silica is a primary carrier structures of the plant, it is with clay, that we can see the 'good heart' bringing qualities, in plants. The key to clays role in stimulating the upward Spirit activity can be found in the role of Silica's intimate partner in clay, Aluminum. While aluminum silicate's are found in nature, something further has happened to these two to form the lattice structure of clay. We know that it takes tremendous energy to separate them in their clay bond, which suggests tremendous energy has gone into clays formation. Clay forms in shallow water environments, and the energy source is most likely the Sun beaming down upon the Earth, everyday. We do not talk much about Aluminum in agriculture, as like Silica there is plenty of it about, however we have recognised the important role Aluminum's smaller brother Boron has in plant growth. Boron is essential for the upward flow of plant sap. If it is in short supply we see the plant wilts through the middle of the day. Boron also has a role with Fluorine in the solubilisation of Silica, so it can become available to life processes. This relationship is seen in glass making. Similarly Aluminum is a very 'softening' element. As an amphoteric element it can act as a acid or a base depending on who it is related to, and shows a similar tendency with Calcium and Silica, moving between both with equal ease. Dr Hauschka talks of its mediating role in the plant between the blossom and the root, when it 'carries the Earth forces upwards from the root, and the sun forces of the blossom to the root.' Which is a mimic of the job done by Boron with regard to physical nutrients carried in the sap. The softening qualities of Aluminum upon the rigid crystallized Silica, can be seen further in clays ability to be modeled by outside forces. With the help of Water Air and Fire, clay becomes very useful pottery. Where the Silica processes dominate the Aluminum we have the development of precious gems like Ruby, Sapphire and Emeralds, along with others. While where life moves towards the Lime pole we form bricks and build houses. Without adequate Boron and Fluorine, Silica substance will not be available to the plant. Similarly Aluminum with its bond to Silica in clays, helps make the Spirit in nature active, and we see it in strong upright Spirit carrying qualities in the upward moving force stream.

As every gardener knows, clay soil by itself will not grow good plants. It is only when humus, sand and missing cations and anions are added, that plants grow well. I can speak with some experience of clays activity, as I have had a garden on volcanic pumice ash for the last 10 years. Therefore there is very low humus, low cation exchange capacity and no clay present in the native soil. Initially I added plenty of compost and other mulch materials and added the appropriate range of minerals indicated by soil testing. The first two years saw good growth, as would be expected of a new garden, however in year three everything stood still and looked 'weak'. I tried more Sulphur, given we have 2000mls of rain a year, and various liquid manures, however nothing really did it. Soil tests were fine and there was plenty of organic activity, but not good growth. Eventually I began applying crumpled yellow clay, obtained from some 100kms away. I used approximately a 6 inch ball of clay to 10 sq mts of bed. After the next rain that section of the garden took off and 'good hearty' vegetables began to grow. 7 years on and the garden is still producing very good crops of vegetables. Capsicums are a very good indicator of the clay function. If it is not strong enough the capsicums will not size up. They set, and stay small and are liable to rot before ripening. Applications of clay both as a substance and as a homeopathic spray, have seen the fruit immediately begin to size and ripen fully, without rots. It is as if the tourniquet is released and the upward sap stream pushes right through the plant and out the fruit.

Phosphorus

Another element the Spirit uses in its journey through matter is Phosphorus. Dr Hauschka talks of its polarity with Aluminum, in the Geosphere, while my circular Periodic Table also identifies Aluminum and Phosphorus, mediated by Silica, as the elements of the Nerve Sense system and its companion the soil.

Unlike Aluminum, Phosphorus is very finely distributed over the Earth, although life processes find access to it from decaying plants and from deposits of bird manure, however healthy soil fungal biology is needed for sufficient plant uptake of phosphorus from organic sources. While Aluminum is a stable element due to forming very strong bonds with other elements, Phosphorus can self ignite if left open to the air, giving off a very bright cold light. Like Aluminum it is amphoteric, and moves easily between Calcium and Silica. Phosphorus proteins make up our nervous system, which we find very active in our skin, while its association to Calcium is found at our center through the formation of Calcium Phosphate in our bones. It is also important in energy generation through the processing of carbohydrates via the ATP process which develops during photosynthesis, and provides energy for many other processes. It is Phosphorus that allows us to bring our picture consciousness into clear reflective thoughts, in our front brain. Here Phosphorus helps Silica incarnate our spirit, so we can achieve self consciousness. Dr Steiner refers to Phosphorus as a dissolving agent that works in our brain. This process continually dissolves the 'earthing' processes of Calcium and Aluminum. Where this process is too weak we see Aluminum condensing and filling our brains, with various dementia states being the result. The Spirit can not stay incarnated when its Earthly companion is 'hogging the limelight'.

In plants, Phosphorus is an essential element, that performs several vital functions that can be seen as 'spirit functions'. It is essential as part of several central plant structure compounds, and as a catalysis in the conversion of several biochemical reactions in plants. It is pivotal in capturing and converting the sun's energy into useful plant compounds, and is a vital part of DNA, the genetic carrier of 'memory' and form of all of life. It is also part of RNA, the compound that interprets the DNA code, so proteins and other compounds essential for plant form can be developed. The structure of both DNA and RNA are held together by phosphorus bonds.

It is also seen as 'the usher at the movies', through its role in the moving various minerals to where they are needed in the plant. In this action we see it is a World Astral element, carrying out the will of its master.

Sulphur

Sulphur (S) also carries the footprints of the Spirit, but in a more internal way than Phosphorus (P). While P is a World Astral element, Sulphur is the element of the Internal Astral arm, both in the Physical sphere. We also see in Sulphur character that it is on the anion side of the manifest etheric arm.

S is found in many more forms and internal functions than P, as seen it its importance in all protein formation. S is changeable enough to appear in 6 or so different forms, which can change from one to the other through heating. Unlike P, which works in 2 forms (elemental P and Phosphoric acid). S is inflammable in the air, and when it is burnt it has a very dark hot flame. It combines easily with Hydrogen to form the unforgettable rotten egg smell of Hydrogen Sulphide. In all its many combinations it provides heat to biological processes, which provides the images of being a Spirit carrier however "Sulphur is a uniting force that promotes cosmic essences to work together in building up matter — where we need to have more life in our physical processes, with no interference from the activities of the soul. — where an excess of it causes dizziness and dimming of unconsciousness" This last image is the opposite of P, which enhances consciousness. Hauschka goes on the say that S has the role of keeping the soul from incarnating ,so that the purely vegetative forces of the physical and etheric can go about their up building work undisturbed. All these images are talking of how S enters into the physical and etheric realm more deeply than P, and acts as the Spirits facilitator on the shop floor, rather than as the director role played by Phosphorus. In the relationship of Phosphorus to Sulphur we see an image of the difference between the working of a World sphere element and a Internal sphere element.

Halogens

Other elements that carry the Spirit are those of the Halogen group and the noble gases. The halogen group which include **Fluorine, Chlorine and Bromine** exhibit one quality of the Spirit as yet unmentioned. The Spirit through its highly contractive nature is a death force. Left to itself it will consume all the etheric forces available to it and bring life processes to an end. We see this in the contraction of the plant to the seed, which in annual plant brings the plant to an end. The Spirits push for consciousness is generally met by the outward push of the physical and etheric, but if the Spirit is too much, they will be consumed and Hydrogen's secondary task of carrying the Spirit back to the stars will be fulfilled. In the Halogens we see this same quality of drawing all life and oxygen to themselves, and especially so in their acidic form when they accompany Hydrogen. Hydrofluoric acid is strong enough to melt a glass rod. Dr Steiner talks of how fluorine rounds off the expansive processes magnesium brings to teeth, in the formation of enamel. In Phosphorus we see processes directed, with Sulphur we have processes sped up and moved along, while with Chlorine we have them bought to an end. RS talks of the positive role Chlorine plays in our stomach, where Hydrochloric acid brings the etheric activity of the food we consume to an end, in doing so our etheric body can be stimulated through this 'struggle'.

Iodine deserves a special mention, as it plays a very pivotal roll in some 32 biochemical sequences, along with being central in the control of the metabolic system, through being the prime element in good thyroid function. Only small amounts are needed, however given Iodine is deficient in most soils away from the oceans, shortages are not uncommon. "Adult population inhabiting the iodine deficient areas is characterized by a high degree of apathy, reduced mental functioning, lack of physical energy and reduced work output, all contributing to poor quality of life. Iodine deficiency has emerged as a socio-medical problem of vast dimensions associated with physical and mental retardation, neurological disorders, feeble mindedness, low educability, poor performance, social handicaps, dependability and disfigurement

From this quote it is obvious that the Spirit's ability to incarnate is facilitated by appropriate levels of Iodine. I have identified the Halogens as being expressions of the World Spirit, while in the very aloof and self contained nature of the noble gases, we can see the expressions of the Internal Spirit. The various members of these families carry the Spirits activity into the other bodies. Chlorine carries the World Spirit into the Physical body, while Bromine carries it into the Etheric body, while Iodine is the carrier of the World Spirit, into the Astral sphere, all on the manifest Spirit anionic arm. In the circular Periodic Table, (PT) if there is one point where we can say 'God' resides, it would be with the last halogen **Astatine**. It is an extremely rare element whose isotopes have a half life of 8.1 hours, so it is very ethereal. This point in the PT is where the Cosmic Spirit meets with the World Spirit, on the manifest anionic Spirit arm. Hence the most 'Spirit' point. So Iodine being the astral element below Astatine, we can expect Iodine is doing Astatine's bidding. Iodine is Gods head contractor — to use the earlier building analogy. Iodine provides the overall guidance for the metabolic functions of the physical body, and especially the development of many brain functions, the 'home' of the incarnated Spirit.

One fact of chemistry is that biological processes make very little differentiation between any one member of a chemical family. A biological process will accept Chlorine, if it is a high enough density in a solution, as a substitute for Iodine, even though the Chlorine does not have the desired 'wattage' provided by Iodine. Hence the process will be unable to complete its cycle and cell death occurs. Eventually the process will become exhausted and other diseases follow as a result.

In the right proportions these World Spirit elements help guide some processes, however where they become too strong, the World Spirit activity displaces the Internal Spirit, and thus we see a depletion of personal will forces, while the physical hardening processes of the Spirit are enhanced. The present toxic levels of Chlorine and Fluorine we are ingesting via our treated water, are standing behind the will less compliance we are seeing presently in our populations, as the corporate takeover the planet occurs unopposed. These 'lower' halogens are swamping our Phosphorus and Iodine, so the populations can not think straight. A+B no longer = C. The dementia epidemic that comes with these elements is growing everyday, due the Internal Spirit being pushed out, and the Astral body being left to its own devices.

Sodium

Of the four main cations, Sodium, Potassium, Calcium and Magnesium I have come to consider Sodium to be the 'Spirit' element. The Albrecht soil system suggests we want to create the ideal soil mineral balance for microorganisms to thrive. Then we need the cations to be in the ratio of approximately 70% Calcium, 14% Magnesium, 4% Potassium and 1% Sodium and 6% Hydrogen. So sodium is the lowest proportion however it is THE element that controls the movement of water in a system. Too much Sodium and water becomes 'held'. It is the dominant element in the extra-cellular fluid, of most living systems and with Potassium also plays a major role in the functioning of the nerve system.

I see Sodium as the Spirit controller of the Etheric, Potassium brings the Astral into the Etheric and water, Calcium is the stimulator of the Etheric element, while Magnesium helps the Etheric to bind with the Physical.

Trace Elements

In the trace elements the Iron family manifests the Saturn 1 activity into the higher bodies, while the Manganese family are expresses of Saturn 2. We can see these activities in Irons role in bringing about the environment for photosynthesis to occur, thus providing the plan for the primary basis of a plants life, while Manganese is an essential element for seed formation and maturity.

Noble Gases and Ring 6

The Spirit can also be 'collected' from the Noble Gases, and from all the elements of the 6th and 7th layers of the Periodic Table see (1) The Noble Gases and the layer 6 elements have a special affinity for the Internal Spirit activities while the Actinides work with the World and Cosmic Spirit activities.

The Spirit and the BD preparations

All of the above elements can be used to influence some activity the Spirit is involved in. Within Biodynamic Agriculture we are also given a few preparations.

Horn Silica (501) is the primary carrier of the Cosmic stream and its activity can be specialised and directed through particular chooses of homeopathic potency, and by the companion chemical elements one might choose to combine it with. While it is a primary strengthener of the nerve sense processes, especially when used in the afternoon, it is also seen to incarnate the light processes through the metabolic region, when applied in the morning. On the one hand it will strengthen the plants structure, on the other it will often force a plant into flowering, and bring great fruit quality.

Valerian (507) is accepted by most BD folk as the Saturn preparation. The relationship it has to the 'phosphoric substance' and the warming influence it brings to plants both indicate this preparation is helping to strengthen the Spirits activity in plants. In some plants this preparation can encourage movement to seed formation, in others, such as trees, it shows an increase in flowering, by often bringing it forward, and shortening the flowering period.

Dandelion, (506) encourages the Spirit and Astrality to work together, and in turn incarnate both more strongly into the physical body. RS comments on it bringing more sensitivity to the plants to receive what they need from their environment. It has also been seen to help with fruit set and fruit sizing and flavour by encouraging the Jupiter 2 processes of alkaloid and oils development in plants.

The specific effect of these preparations can be enhanced further by using them in a particular warmth constellation and by the choice of homeopathic potency.

Cow Horn Manure (500) needs a comment, because of its time in the ground during the fructification and crystallization periods. Given this is the time when the soil is most alive with cosmic forces and earthly substances, it is fair to assume that this preparation will carry some of both of these activities. Cow manure is a 'humus' substance of the metabolic sphere, rather than the nerve sense sphere, and when applied to the soil it has a very enlarging 'Earthly' effect on plant growth. When used in excess it makes for big leaves and pushes off the flowering process of plant, allowing some plants to remain in their leafing mode for a longer period. This indicates its is more related to the Earthly Substance processes than the Cosmic Forces. In clay soils this may not be such an issue, however in sandy soils a tendency to rotting diseases could be enhanced. So while the 'strength' of plants grown with 500 suggests good cosmic forces, these other indications suggest it is a very secondary influence, to the dominant Earthly Substance processes.

The Preparations and the Constellations

The question of working with the Constellations, to stimulate the Spirit activity further, can be answered from two standpoints. Both of which should be very useful, even though they focus the Spirit in two different directions.

The first will be considered 'traditional' , while the second is one of my innovations based upon RS, Lievegoed and Kolisko's suggestions

Aries	Sagittarius	Leo
Cardinal	Mutable	Fixed
Metabolic	Rhythmic	Nerve Sense
Flower	Leaves	Root
Mars	Jupiter	Sun /Saturn
Astrality	< >	Spirit
Flower	Fruit flesh	Seed
1-10	10-20	20-30

From the Thun sidereal approach to the Zodiac, it would be expected that if we wish to encourage the Spirit warmth axis of our creation we would work with the Fire constellation, Aries, Sagittarius and Leo. So we can suggest Valerian and Silica applied in one of these constellations will stimulate the activity of the Spirit in nature.

If we use the chart I presented in 'Biodynamics Decoded' for defining the effect of the constellations, we can see Aries is the Cardinal / Metabolic constellation, while Sagittarius is the Mutable / Rhythmic constellations, with Leo being the Fixed / Nerve Sense constellation.

Likewise Aries is ruled by Mars, Sagittarius by Jupiter and Leo is ruled by the Sun, whose polarity is Saturn. So if we are looking for the Spirit part of this story we would go with Leo, Aries is the Astrality, while Sagittarius has the job of join these two together and bring them closer to the Physical body. Add the potency choice and you will have a 'spell' worth applying.

Jupiter 1	Saturn 1	Saturn 2	Jupiter 2
Pisces	Aquarius	Capricorn	Sagittarius
Cobalt	Iron	Manganese	Chromium
506	507	507	506
	Flower	Leaves	Root
	1-10	10-20	20-30

It can be further identified that as this reference system is based up the 3 fold modes, we are working with the Physical sphere, and focusing upon the element of fire, behind these constellations we are reaching the 4 fold Etheric sphere. So I suspect this method would be good for controlling the growth of the plant, within that season.

In the second method of working with the Constellations, we look at the references that are developed from following the planets, and in particular the image of the zodiac arising from the dual constellational rulership of the planets. The details of double planet patterns, were developed further for BD by Dr Lievegoed. In 'Energetic Activities' I completed the Zodiacal relationships for the double planetary activity. This work arose from the question of how we might be able to focus the existing BD preps into their primary or secondary planetary functions. Thus this relationship to the zodiac became one possible solution. When looking at the Valerian preparation the two constellations are Saturn 1 being Aquarius, while Saturn 2 is Capricorn. Similarly if we wished to emphasis the Jupiter preparation Dandelion, we would use Pisces for the Jupiter 1 constellation and Sagittarius to emphasis the Jupiter 2 process.

To further enhance these actions we might consider which trace elements are related to these planetary activities. In the case Of Saturn 1 it is Iron, Saturn 2, Manganese, Jupiter 1 is Cobalt and Jupiter 2 is Chromium. These elements are on the Etheric ring of the Periodic Table, and so will focus the Spirits activity within the Etheric sphere. To work on the Astral and Spirit spheres more strongly, the appropriate 'brothers' of the elements mentioned can be used. Potency choice for any of your remedies can further refine their focus to a particular physical system.

Given these references are arising from the planetary sphere, these remedies can be expected to work more deeply into the plants Astral qualities and processes, as described by Dr Lievegoed, and may well show up significant changes in subsequent generations.

While I have explored both these reference systems, and feel comfortable with my suggestions, this is a huge avenue of research and so much more needs to be done before I can speak with complete confidence in this approach. It worth putting here as hints of a way forward.

Epilogue

It has been fun being with this essay, and I trust there is enough here to emphasis that the Spirit is a cornerstone part of the Biodynamic 'game', and that it needs to be continually included in ones practical considerations. If there is no Spirit, there is no plan, no direction, no upward growth, and we see all the other processes run amok, leading to all manner of viral, bacterial and fungal diseases.

Biodynamics, is a world view based on the four primary parts, sourced from our environment working into three Physical organisations. If a BD story, does not place itself within this context, it needs to be questioned very seriously.

References

1) Glenological Chemistry p 86, 98 , Glen Atkinson (GA)
2) The Chemical Basis of Medical Climatology, Prof. G Piccardi ISBN 039807049-0
3) See Case Studies – www.garudabd.org— coriander
4) Nature Of substance p 133, Dr Hauschka
5) Nature Of substance p 136, Dr Hauschka
6) Glenological Chemistry p 147, , GA
7) Fundamentals of Anthroposophical Medicine. Pg 39. Dr Steiner
8) Spiritual Science and Medicine p 142 Dr Steiner
9) Energetic Activities pg. 82 GA
10) See 'Case Studies' at www.garudabd.org — capsicums
11) 'The Ten Bulls'
12) The Anthroposophical Approach to Medicine, Part 2, pg. 273, Hausmann and Wolff
13) Physiology and Therapeutics, Lec 4, pg. 3, Dr Steiner
14) The Working of the Planets and the Life Processes in Man and Earth— Introduction,   Dr Lievegoed
15) The Twelve Groups of Animals , E Kolisko
16) See the diagram in the text
17) Biodynamics Decoded pg. 107 , GA
18) Energetic Activities , Part 2, GA
19) Glenological Chemistry, pg. 119, GA
20) The Etheric Formative Forces , Collected Essays , GA
21) The Four Ethers, E. Marti
22) Phenomenology, Tom van Gelder
23) Glenological Rosetta Stone— picture, GA
24) Biodynamic Questions, Astrological Answers, GA
25) Iodine, Iodine metabolism and Iodine deficiency disorders revisited http://www.ncbi.nlm.nih.gov/pmc/articles/PMC3063534/   Indian journal of endocrinology and metabolism
26) Nature Of substance p 152, Dr Hauschka
27) Glenological Chemistry p 133, , GA
28) Lecture 7 - Man as Symphony of the Creative Word
29) Lecture 5 - Cosmic Workings in Earth and Man
30) The Spirit in the realm of plants 8 dec 1910
31) Agriculture Course GA 327 Lec 8
32) Human Questions Cosmic Answers, 2 Jul. 1922 GA213
33) The Groups souls of Animals Plants and Minerals, 2 Feb. 1908

Cell Salts
Glen Atkinson

© Garuda Consultants ltd
16 April 2025

Another avenue of investigation are the 12 Biochemic Cell Salts, that are used as human remedies. These salts were discovered by Dr Schuessler some 150 years ago. They are all found in the human cells and are noted to be related to specific disease representations, when they are in short supply. The salts contain most of the main elements used in agriculture. The 12 salts are combinations of Ca, Mg, K, Na, P, S, O, Cl, F, Si, Fe. It seems appropriate to explore how these readily available salts are of use to horticulture and especially biodynamic agriculture.

There are various authors addressing this on the internet, however most of this information about the cell salts is gained from external observation. Schuessler ashed dead bodies of known diseases and then examined the deficiencies. This has not been done with plants, as far as I know, nevertheless trial and error has produced a guide for their use.

Can we find the energetic underpinnings of these remedies?

The Cell Salts and the Zodiac

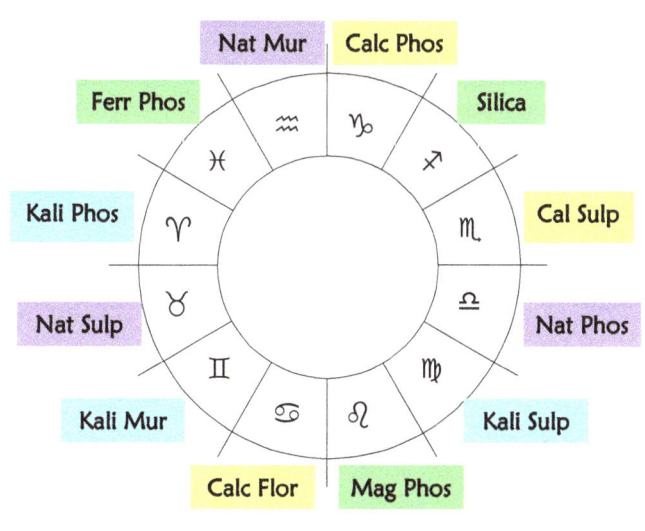

Traditional Zodiacal References
Perry & Carey 1932

There has been zodiacal references, given for the cell salts, for a very long time. It appears that there is a general consensus on these relationships, and they all appear to track back to the book by Perry & Carey " The Zodiac and the Salts of Salvation", first published in 1932. In the intervening years many practitioners have found that particular zodiac signs have deficiencies of the companion cell salt. However generally what one reads of these associations, they are based upon peoples interpretations of the sign, (often with a northern hemisphere seasonal references) and how this correlates to the characteristics of the cell salts. These traditional relationships are shown in the diagram on the previous page.

The Cell Salt Elements

Anions

The anions can be categorized into three groups

Halogens   — F, Cl
Oxides   — S, O
Phosphates   — P

Hauschka (26) provides good pictures of each of these, The Halogens bring processes to an end, as shown for example in the rounding off of glass when dipped into it. Halogens are very acidic and as they bring in the spirit they 'kill' things. They stop them. Hence the **Halogen elements can be related to the processes of the metabolic system.** Hydrochloric acid is only found in the digestion and its role is to totally breakdown the foods and their forces, to make them possible for our forces to 'digest' them. Being related to the Metabolism they must be related to the Cardinal mode in Astrology.

The Oxide / Sulphates are elements that facilitate living processes. S acts as the oil of the spiritual bodies interaction in protein, while O acts as the carrier of the etheric which in itself acts as the component that allows the other bodies to come into Living forms. These are the facilitators of the Mutable elements.

Phosphate in plant growth acts as the 'usher' of other elements to their place. It gives the plant its direction. This makes it the obvious directional element. In the Human this direction ability comes from the nerve sense system and thus relates Phosphorus to the Fixed Element.

Halogens	F, Cl	Cardinal
Oxides	S, O	Mutable
Phosphates	P	Fixed

Cations

As I want to relate the cell salt elements to the four spiritual activities it makes sense to associate them to the four primary cations. There are several orderings of this association. There are six examples from page 174. I have collected to date. What is your suggestions?

At present I see them as

Spirit	Sodium	Na
Astral	Potassium	K
Etheric	Calcium	Ca
Physical	Magnesium	Mg

The relationship of Calcium to Magnesium needs careful consideration

On my Periodic Table, Calcium is the most etheric element, being Cosmic Etheric, Internal Etheric, Manifest Etheric, (see pg. 197) and is described in the Ag course as the element that holds and strengthens the etheric activity into living processes. It is interesting that once astronauts go beyond the Earth atmosphere (World Etheric) they have trouble maintaining the working of Calcium. However in the oak bark preparation and in its ability to 'suck' nitrogen to itself and hold elements it is providing the image of drawing the etheric into the physical organism. Thus leading to an enlivened physical body.

Magnesium is on the Cosmic Physical, Internal Etheric, Manifest Etheric physical element. (see pg. 197) It is on the Cosmic Physical ring and is described in Apop literature as bringing the Etheric into the Physical. In soils its increased presence binds the soil together into rock. Being in the 3rd ring of the periodic table, it still has very strong electron bonding and will hold itself together very strongly. However we see it activating the etheric activity in the expansion of teeth, the relief of cramps and the expansion of plant growth through photosynthesis. Hence we have the picture of the etheric being activated from the physical body outwards, somewhat opposite to Calcium.

Potassium and Sodium present a little problem. I feel K is the light bearer of the cations, it is associated with building the framework of the stems and its pivotal role in photosynthesis, by controlling stomatal conductance, and in the development of proteins and sugars. All Mars 2 and Jupiter 2 activities suggesting the working of the Astral and Spirit. It maintains the water pressure in the cell by directing water into the cell (astral directing etheric activity).

Sodium strikes me as the 'hard' element of these four. It controls the water movement within the plant by holding it to itself. On the Periodic Table it is on the Cosmic Physical ring, which can seem a contradiction however Physical and Spirit are a creative polarity and on the Periodic Table there are no 'spiritual' cations, until we come to the Cosmic Physical/ third level of the Periodic Table activity with the introduction of the Trace Elements, some of which work with the Astral and Spirit. So where would we find

		CARDINAL	MUTABLE	FIXED
		Cl	S	P
FIRE	Na	♈	♐	♌
		Nat Mur	Nat Sulp	Nat Phos
AIR	K	♎	♊	♒
		Kali Mur	Kali Sulp	Kali Phos
WATER	Ca	♋	♓	♏
		Calc Flour	Calc Sulp	Calc Phos
EARTH	Mg	♑	♍	♉
		Mag Mur	Mag Sulp	Mag Phos
		Silica	Fe Pos	

ASTROLOGY, Elements & Modes, Atkinson Cations & Anions, **Cell Salts**

the spirit working in these cations, except in the physical elements. Sitting next to Mg, the physical body element in this set, would suggest Na as Spirit carrier has some appropriateness. Finally Na has a strong relationship to Chloride in salt, indicating Na again having a strong 'spirit' affinity.

Given these associations it is possible to make a cross reference of the Astrological indicators and the cell salts elements. This ordering thus brings the cation elements into a trine relationship

	CARDINAL	MUTABLE	FIXED
FIRE	♈ Kali Phos	♐ Silica	♌ Mag Phos
AIR	♎ Nat Phos	♊ Kali Mur	♒ Nat Mur
WATER	♋ Calc Flor	♓ Ferr Phos	♏ Calc Sulp
EARTH	♑ Calc Phos	♍ Kali Sulp	♉ Nat Sulp

ASTROLOGY, Traditional Cell Salts

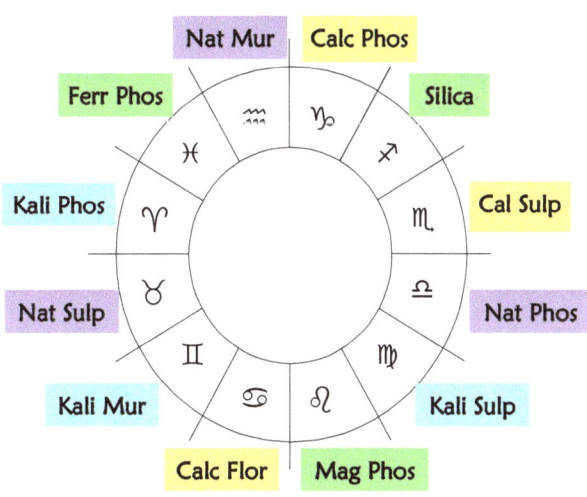

Traditional Cell Salts to Zodiac

with each other, which is different to that found in the tradition associations where the trine square quinqunx relationships become evident. This ordering thus provides an alternative order to the traditional order. This does not make either right or wrong it just provides two streams of future research options.

The Chemical Elements

It would make some sense to address these elements in the order they appear in atomic weight, and thus as they 'descend' down the spiral into denser material forms. However for the sake of finding relationships between the elements, I will address them in the order of the Arms they are on, as several elements share Arms. Na and K are on the World Etheric arm, while Ca and Mg are on the Internal Etheric arm. Highlighting all are active with the Etheric is some way.

Sodium   Na
Chemical Data: 11   22.990
Layer: 3— Cosmic Physical
Gyro Arm: World Etheric
Steiner Arm: Manifest Astral +
Ag. Course Name: Cosmic Calcium
Planetary Ring: Sun
Spiritual Activity: Cosmic Physical forces work upon the World Etheric into the Manifest Astral Cations.

In Sodium we see the 'alkali' ability to form colloids and form surfaces , while at the same time controlling the activity of water within a system. In the forming of colloids we see the tendency to 'individualise' something from the whole, hence we have a 'spirit' activity present.

In its controlling of water we see that it orders, restricts and moves water within a system, again an action that can be a symbol of the spirits influence upon the etheric.

It acts as an electrolyte and energise the flow of electrical energy within an organism, while in its physical nature, in air it has to be kept in oil to inhibit it from spontaneously combusting. While RS has stated that without salt there would be no thought. Sodium and Potassium are both elements necessary for the conveying of messages within the nervous system as the chemicals at the nerve synapses. Sodium is more dominant in the animal kingdom, and poisonous in the plant kingdom.

Sodium seems to encourage crop yields and in specific cases it acts as an antidoting agent against various toxic salts. It may act as a partial substitute for potassium deficiencies. Excess may cause plant toxicity or induce deficiencies of other elements. If sodium predominates in the solution calcium and magnesium may be affected.

All these things are indications of the activity of the Spirit.

Potassium   K
Chemical Data: 19   39.098
Layer: 4 Cosmic Etheric body
Gyro Arm: World Etheric
Steiner Arm: Manifest Astral +
Ag. Course Name: Free Water / Chemical ether
Planetary Ring: Jupiter
Spiritual Activity: Cosmic Etheric into the World Etheric into the manifest astral cations

Again an alkali so again the ability to form colloids and surfaces within amorphous masses. However we see it is necessary for the formation of stalks and fruit within plants. It appears to have a stronger relationship to light and its functioning in plants and while it is a 'bigger' element than Sodium it appears to work as the 'younger' brother to Na.

This association with light suggests its relationship to the Astrality.

"Together with nitrogen and phosphorous, potassium is one of the essential macro minerals for plant survival. Its presence is of great importance for soil health, plant growth and animal nutrition. Its primary function in the plant is its role in the maintenance of osmotic pressure and cell size, thereby influencing photosynthesis and energy production as well as stomatal opening and carbon dioxide supply, plant turgor and translocation of nutrients. As such, the element is required in relatively large proportions by the growing plant.

The consequences of low potassium levels are apparent in a variety of symptoms: restricted growth, reduced flowering, lower yields and lower quality produce.

High water soluble levels of potassium cause damage to germinating seedlings, inhibits the uptake of other minerals and reduces the quality of the crop. " (Lenn tech)

"Potassium has two roles in the functioning of plant cells. First, it has an irreplaceable part to play in the activation of enzymes which are fundamental to metabolic processes, especially the production of proteins and sugars. Only small amounts of potassium are required for this biochemical function.

Second, potassium is the "plant-preferred" ion for maintaining the water content and hence the turgor (rigidity) of each cell, a biophysical role. A large concentration of potassium in the cell sap (i.e. the liquid inside the cell) creates conditions that cause water to move into the cell (osmosis) through the porous cell wall .

Turgid cells maintain the leaf's vigour so that photosynthesis proceeds efficiently.

Photosynthesis is the process by which plants harvest the energy of sunlight to produce sugars. These sugars contain carbon derived from the carbon dioxide in the air that has entered the leaf through the stomata, tiny openings mainly on the underside of the leaf. These tiny openings are surrounded by "guard cells" and it is only while they are turgid that the stomata remain open and carbon dioxide can pass through into the leaf. But most of the water transpired by the plant is lost through the stomata when they are open. Thus, if there is a water deficit, the plant needs to close the stomata to conserve water. The plant controls the opening /closing of the stomata by regulating the concentration of potassium in the guard cells. A large concentration of potassium ensures turgid cells and open stomata. When the potassium in the guard cells is lowered, they become limp and the stomata close.

Stomata (plant leaf pores) natural size 0.055 mm

Potassium ensures the turgor, or rigidity of plant cells. While the guard cells surrounding the stomata are rigid the stomata remain open, allowing carbon dioxide to pass into the leaf where the carbon is converted to sugars.

A high osmotic potential in plant cells is also needed to ensure the movement through the plant of nutrients required for growth, and the sugars produced by photosynthesis, for example, the transport of sugar to grains, beet roots, tubers, and fruit. By maintaining the salt concentration in the cell sap, potassium helps plants combat the adverse effects of drought and frost damage and insect and disease attack. It also improves fruit quality (Box 2) and the oil content of many oil-producing crops."
www.efma.org

"The rate of photosynthesis drops sharply when plants are K deficient. Too much N with too little K can result in a back-up of the protein building blocks, set the stage for disease problems, reduce production of carbohydrates, reduce fruiting, and increase fruit creasing, plugging and drop. A shortage of K can result in lost crop yield and quality. Moderately low plant K concentrations will cause a general reduction in growth without visual deficiency symptoms. The onset of visual deficiency symptoms means that production has already been seriously impaired."
http://edis.ifas.ufl.edu/SS419

Magnesium   Mg

Chemical Data: 12   24.305
Layer: 3 Physical body
Gyro Arm: Internalised Etheric
Steiner Arm: Etheric +
Ag. Course Name: Cosmic Substance
Planetary Ring: Sun
Spiritual Activity: Physical body's influence on the Etheric body

Mg is described by RS as the element that **anchors the etheric body into the physical.** He talks of its process in teeth development as being expansive and one that needs to be checked by the fluorine, otherwise it would have for very big teeth. There is also the image that without adequate Fluorine an individuals extremities become elongated, suggesting this outward movement of Magnesium being very strong.

"This balance impresses the primitive human form that originates in the cosmos into earthly etheric events. This happens in the field of tension between magnesium, which radiates from within outward, and fluorine, which delimits and mineralizes from without." Med Lec 1920

I appreciate in soil science that high Magnesium in a soil causes it to become very sticky, locked up and hard. We can see this in it being an element of the physical ring of the PT. Yet Mg also has a specific relationship to light and its functioning especially in the plant kingdom. Only when there is enough Mg and sufficient light, is it possible for photosynthesis to occur and for the plant to expand its growth, thus allowing the etheric body to become active. Magnesium is an important mineral in all biochemical reactions within the body and in this way mimics its opposite Sulphur. They seem a pair that works internally to allow life and the etheric body to express itself. Both with a gesture of moving from inside outwards.

We see in the recommendations of the cell salts that MgPhos is used to combat cramps. Cramps are caused by the astral body becoming to active within the physical, due to a build up in Uric Acid in the blood and muscles, and thus constraining the muscles. We see a similar thing occurring with heart palpations. The K levels rise in the body enhancing the quick astral pulse thus disturbing the natural 4: 1 balance between the lungs and the heart. Mg is used to restore the 'balance' with the K.

In both these examples we have an image of the etheric stimulating Mg pushing back the Astral K, and thus restoring balance.

Calcium   Ca

Chemical Data: 20   40.078
Layer: 4 Etheric body
Gyro Arm: Internalised Etheric body
Steiner Arm: Etheric +
Ag. Course Name: Bound Chemical ether
Planetary Ring: Jupiter
Spiritual Activity: Internalised etheric formative forces

Calcium also provides a mixed picture as it sits on the Cosmic Etheric , Internal Etheric, Manifest Etheric position of the Periodic Table, and can be seen to lead to the expansion of growth in plants. However its action as an element suggests that it has a very sucking and attracting nature , similar to a very physical element. It is an element linked to the Earth gravity, as we see it gives us our weight and structure in the form of our bone formation. We also see in it pathological manifestations such as arthritis, the wrongly placed calcium leads to us seizing up and becoming sclerotic. Our life in outer space, is hindered by the improper working of Calcium, when we are away from the Earths gravity. All the Calcium on Earth has come to be concentrated by the deposits left from physical bodies of animals. Be they sea animals or bones from animals and humans. Suggesting its basic activity is associated with the Earth. As calcium has become anchored on the Earth so has the anchoring of the etheric body into the physical been possible.

We can see in the soil we need a base amount of around 70% Ca to have a neutral enough pH, to have healthy biological and plant growth, in our modern agriculture. Similarly high Ca levels are needed in the human to have a basic state of health.

We also know that Ca acts as a drawing and attracting element for elements such as Nitrogen and Phosphorus, helping to stabilise and even fix them. If ever Calcium is breathed in, it has a 'take your breath away' effect indicating a drawing towards and holding tendency of the physical body.

With the biodynamic preparations we see the Oak Bark preparation, which is high in mineral calcium, is used to draw the etheric body to the physical.

Overall Calcium's importance in the stimulation of the Etheric activity suggests it is an element of Water, the other essential ingredient for Life.

Silica   Si

Chemical Data: 14   28.086
Layer: 3 Physical bodies
Gyro Arm: Internalised Physical body
Steiner Arm: Physical +
Ag. Course Name: Earthly Matter
Planetary Ring: Sun
Spiritual Activity: Carrier of the physical formative forces

The energetic activities all have physical carriers. Each primary activity has several chemical agents they can work through. (see Glenological Chemistry). In the agriculture course RS mentions how the Astral and Spirit activities are carried and enhanced by Silica , while the Physical and Etheric activities are carried on the Calcium elements. We have seen how the kidneys have a particular relationship to the workings of the Astrality and so this function will depend upon the working of Silica. There are however two distinct forms of

Silica. We have mineral quartz Silica often taking the form of a crystal, and then we have Silica interacting with life, through its relationship with the Etheric carrying water, in the form of Silicic acid. H4SiO4. This suggest that one atom of Silica can hold two molecules of water to itself. In doing so the Silica forces become more available to life processes.

These two forms of Silica make a polarity of activity with each other. It is clear in the medical lectures that mineral Quartz in the form of a crystal, is the foundation of the nerve sense system, and using it as a remedy has the effect of pulling the Spirit and Astrality strongly inwards, from the Head downwards. It also has the effect, when used in low potencies of drawing infections to the surface, so that draining can occur, often through a 'boil bursting'

Silicic acid, H4SiO4, occurs, in many places in nature, most notably in the oceans. In this form it is also found within living processes, having been picked up by the living Etheric activity. It has a more internal working action than Quartz. When used as a remedy for infection, it will stimulate the body to digest the infection from within. In this way it works with the metabolic processes rather than the N/S.

Phosphorus   P

Chemical Data: 15   30.974
Layer: 3 Physical bodies
Gyro Arm: External Astrality
Steiner Arm: Astral —
Ag. Course Name: Terrestrial Silica
Planetary Ring: Sun
Spiritual Activity: World Physical into the Astrality

The usher acts as the transporter element that takes other elements to where they need to be. Thus is the Cardinal / Fixed element. Cardinal in that it is directive, Fixed in that it is the element of the Nerve Sense system, which is related to the head and related to the Fixed pole of the plant. The old Astro BD duality on the modes.

RS comments that P is needed in the brain to stop the natural tendency of the brain to calcify, in this way maintaining a continual state of rickets (softening of the bones) in the brain that we need to allow for thinking to occur. It is the P in the bones that allows for the calcium in the bones to be dissolved into the blood and thus maintaining an even blood pH.

P is also used in the formation of ATP, the basic element of energy use within living processes. So it is a very active and fluid element that gives energy and direction to processes.

Another element the Spirit uses in its journey through matter is Phosphorus. Dr Hauschka talks of its polarity with Aluminum, in the Geosphere (5), while my circular Periodic Table (6, 23) also identifies Aluminum and Phosphorus, mediated by Silica, as the elements of the Nerve Sense system and its companion the soil.

Unlike Aluminum, Phosphorus is very finely distributed over the Earth, although life processes find access to in from decaying plants and from deposits of bird manure, however healthy soil fungal biology is needed for sufficient plant uptake of phosphorus from organic sources. While Aluminum is a stable element due to forming very strong bonds with other elements, Phosphorus can self ignite if left open to the air, giving off a very bright cold light. Like Aluminum it is amphoteric, and moves easily between Calcium and Silica. Phosphorus proteins make up our nervous system, which we find very active in our skin, while its association to Calcium is found at our center through the formation of Calcium Phosphate in our bones. It is also important in energy generation through the processing of carbohydrates via the ATP process which develops during photosynthesis, and provides energy for many other processes. It is Phosphorus that allows us to bring our picture consciousness into clear reflective thoughts, in our front brain. Here Phosphorus helps Silica incarnate our spirit, so we can achieve self consciousness. Dr Steiner (13) refers to Phosphorus as a dissolving agent that works in our brain. This process continually dissolves the 'earthing' processes of Calcium and Aluminum. Where this process is too weak we see Aluminum condensing and filling our brains, with various dementia states being the result. The Spirit can not stay incarnated when its Earthly companion is 'hogging the limelight'.

In plants, Phosphorus is an essential element, that performs several vital functions that can be seen as 'spirit functions'. It is essential as part of several central plant structure compounds, and as a catalysis in the conversion of several biochemical reactions in plants. It is pivotal in capturing and converting the sun's energy into useful plant compounds, and is a vital part of DNA, the genetic carrier of 'memory' and form of all of life. It is also part of RNA, the compound that interprets the DNA code, so proteins and other compounds essential for plant form can be developed. The structure of both DNA and RNA are held together by phosphorus bonds.

It is also seen as 'the usher at the movies', through its role in the moving various minerals to where they are needed in the plant. In this action we see it is a World Astral element, carrying out the will of its master.

Oxygen   O

Chemical Data: 8   15.999
Layer: Two - Cosmic Substance
Gyro Arm: Internalised Astral
Steiner Arm: Etheric —
Ag. Course Name: Sand
Planetary Ring: Mercury
Spiritual Activity: Carrier of Etheric activity

All of the free oxygen on Earth has been released after going through photosynthesis within a living plant. This process is fueled by the Sun forces. As it built up in the atmosphere, so the proliferation of species could occur. It is however the great stabiliser, in its desire to oxidise any raw chemical element. It brings its Life to Death forces and keeps attaching to it until it is stable.

It is the first element on the Internal Astral arm, suggesting its Anionic and highly mobile nature, however it is on the Manifest Etheric Arm, which indicates that it brings its Astral movement to aid the Etheric in its journey into Life. Thus it is the primary element for carrying the etheric into living processes, and moving other substances about while doing so.

In the Earth Oxygen is stable and can be said to be in a 'Sal' state. In the oceans where there is an immense amount of oxygen as water, it is in its 'Mercury', while in the Atmosphere it is in its 'Sulf', and highly reactive state. Remember there is no fire without Oxygen.

Sulphur   S

Chemical Data: 16   32.066

Layer: 3 Physical bodies

Gyro Arm: Internalised Astral body

Steiner Arm: Etheric —

Ag. Course Name: Cosmic Matter

Planetary Ring: Sun

Spiritual Activity: Facilitator of physical activities in the Astral body

Sulphur works from inside out, and moves things along. In the Bee lectures it is suggested to be the Mutable Mode of the Rhythmic system and the facilitator of biochemical processes

Sulphur (S) also carries the footprints of the Spirit, but in a more internal way than Phosphorus (P). While P is a World Astral element, S is the element of the Internal Astral arm, both in the Physical sphere. We also see in S character that it is on the anion side of the manifest etheric arm.

S is found in many more forms and internal functions than P, as seen it its importance in all protein formation. S is changeable enough to appear in 6 or so different forms, which can change from one to the other through heating. Unlike P, which works in 2 forms (elemental P and Phosphoric acid). S is inflammable in the air, and when it is burnt it has a very dark hot flame. It combines easily with Hydrogen to form the unforgettable rotten egg smell of Hydrogen Sulphide. In all its many combinations it provides heat to biological processes, which provides the images of being a Spirit carrier however "Sulphur is a uniting force that promotes cosmic essences to work together in building up matter — where we need to have more life in our physical processes, with no interference from the activities of the soul.— where an excess of it causes dizziness and dimming of unconsciousness" (26) This last image is the opposite of P, which enhances consciousness. Hauschka goes on the say that S has the role of keeping the soul from incarnating ,so that the purely vegetative forces of the physical and etheric can go about their up building work undisturbed. All these images are talking of how S enters into the physical and etheric realm more deeply than P, and acts as the Spirits facilitator on the shop floor, rather than as the director role played by Phosphorus. In the relationship of Phosphorus to Sulphur we see an image of the difference between the working of a World sphere element and a Internal sphere element.

Fluorine   F

Chemical Data: 9   18.998
Layer: Two - Cosmic Substance
Gyro Arm: External Spirit
Steiner Arm: Spirit —
Ag. Course Name: Hydrogen
Planetary Ring: Mercury
Spiritual Activity: Anchor of the world spirit into matter

Other elements that carry the Spirit are those of the Halogen group and the noble gases. (1) The halogen group which include **Fluorine, Chlorine and Bromine** exhibit one quality of the Spirit as yet unmentioned. The Spirit through its highly contractive nature is a death force. Left to itself it will consume all the etheric forces available to it and bring life processes to an end. We see this in the contraction of the plant to the seed, which in annual plant brings the plant to an end. The Spirits push for consciousness is generally met by the outward push of the physical and etheric, but if the Spirit is too much, they will be consumed and Hydrogen's secondary task of carrying the Spirit back to the stars will be fulfilled. In the Halogens we see this same quality of drawing all life and oxygen to themselves, and especially so in their acidic form when they accompany Hydrogen. Hydrofluoric acid is strong enough to melt a glass rod. Dr Steiner talks of how fluorine rounds off the expansive processes magnesium brings to teeth, in the formation of enamel. In Phosphorus we see processes directed, with Sulphur we have processes sped up and moved along, while with Chlorine we have them bought to an end. RS talks of the positive role Chlorine plays in our stomach, where Hydrochloric acid brings the etheric activity of the food we consume to an end, in doing so our etheric body can be stimulated through this 'struggle'.

Chlorine   Cl
Chemical Data: 17   35.453
Layer: 3 Physical bodies
Gyro Arm: World Spirit
Steiner Arm: Spirit —
Ag. Course Name: Cosmic Silica
Planetary Ring: Sun
Spiritual Activity: World Physical FF on the External Spirit

Chlorine is a key energy reactions in plants, specifically the chemical breakdown of water in the presence of sunlight. It also helps control water loss and suppresses disease and infections

(http://www.borax.com/agriculture/files/micro.pioneer.pdf#search=%22Chlorine%20essential%20plant%20element%22)

It is involved with oxygen production in photosynthesis. Chlorine is necessary for osmosis and ionic balance; it also plays a role in photosynthesis. A lack shows in reduced growth, stubby roots, interveinal chlorosis, non succulent tissue (in leafy vegetables)

Photosynthesis is achieved by the plant uptaking water and the sunlight splits the water with the help of Chlorine, into H2 and O. This O is released to the air. The Incoming CO2 joins with the H2 to make C6H12O6 — chlorophyll

Chlorine (Cl) – this element controls water uptake and transpiration. Stimulates photosynthesis and is a major constituent of the anthocyanin molecule. Deficiency – plants wilt easily. Bronze colored leaves with dead or chlorotic spots, stunted roots with club-shaped tips. Toxicity – saline poisoning, small dark leaves, burned margins and wilting. (http://www.hydroponicsbc.com/nutrients.html)

Interferes with P uptake

"Chloride is involved in the evolution of oxygen in the photosynthesis process and is essential for cell division in roots and leaves. Chlorine raises the cell osmotic pressure and affects stomata regulation and increases the hydration of plant tissue. Levels less than 140 ppm are safe for most plants. Chloride sensitive plants may experience tip or marginal leaf burn at concentrations above 20 ppm. Plants with chlorine deficiencies will be pale and suffer wilting. Excesses will cause burning of tips and margins, and bronzing and abscission of the leaves." (http://www.greenair.com/nutrient-properties.htm)

Chlorine is used extensively in organic and inorganic chemistry as an oxidizing agent and in substitution reactions because chlorine often imparts many desired properties in an organic compound when it is substituted for hydrogen (as in synthetic rubber production). It has the highest electron affinity among halides. http://en.wikipedia.org/wiki/Chlorine#Oxidizing_agent

It is found in its highest concentration in the stomach as Hydrochloric acid to dissolve food. Hence it is the Metabolic Anion.

Iron   Fe

Chemical Data: Atomic No -   26,
Atomic Weight - 55.845, solid at 298K
Layer: 4 Etheric body
Gyro Arm: World Spirit / Internalised Spirit
Steiner Arm: Spirit —
Planetary Ring: Saturn
Planetary element: Saturn 1
The spiritual archetype upon which life builds
Major element relative: Bromide (weak)
Venus Pentagram: A
Physical force dominate Spirit - manifesting
Ag. Course Name: Free Fire / Bound Warmth

In Nature: Carries oxygen and ego forces in the blood, helps cosmic weightless elements enter the sphere of gravity, enables us to anchor our personalities in our bodily processes, if low no 'presence of mind', close to Carbon, but has to balance S processes, Fe renders the cyanide processes harmless, has similarities to Pb in it ability to spontaneously combust when ground very fine. Carrier of the forces of embodiment > mummification, ability to absorb and retain formative forces, gives and takes oxygen easily, the breather metal, sensitive to light, negates poisoning processes of As and CN, bivalent form similar to Zn, Fe is light sensitive.

The Saturn 1 transition element in the Cosmic Etheric ring, mirrors what is said above. It suggests the element to allow the Spirit to direct the Etheric activity.

RS often suggest it as a Rhythmic system remedy as in his Bidor remedy, against migraines and digestive issues, Where Silica is used to incarnate the Spirt more strongly into the metabolism , Sulphur is used to activate the metabolism and Iron is used to harmonise these two in their activity.

Functions of iron (Fe)
   * Catalyzes the production of chlorophyll.
   * Involved in some respiratory and photosynthetic enzyme systems.
   * Involved in the reduction of nitrates and sulfates.

Homeopathy: Best adapted to young weakly persons, anæmic and chlorotic, with pseudo-plethora, who flush easily; cold extremities; oversensitivity; worse after any active effort. Weakness from mere speaking or walking though looking strong. Pallor of skin, mucous membranes, face, alternating with flushes. Orgasms of blood to face, chest, head, lungs, etc. Irregular distribution of blood. Pseudo-plethora. Muscles flabby and relaxed.

Spiritual Activity: carries the spirit into the controlling of the etheric activities behind physical processes

Cell Salts through the 3 Stages

As an exercise the traditional cell salts can be moved through the 3 Worlds. Especially the Manifest World organization maybe of interest to explore for which Salt may correspond to a particular time of the Seasons. The accompanying images (left) are facing South.

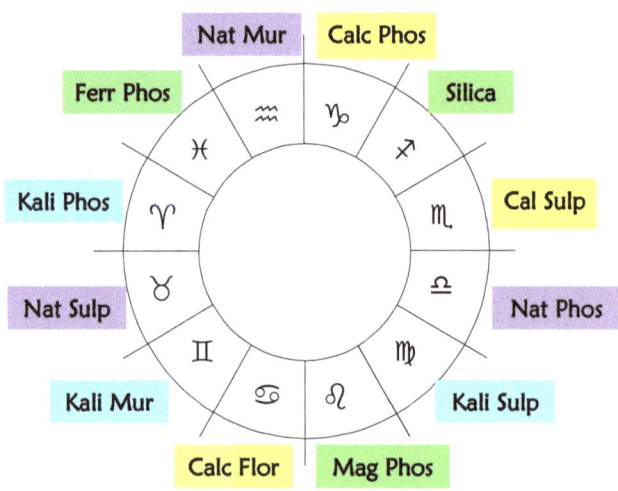

Traditional Cell Salts to Zodiac

	CARDINAL	MUTABLE	FIXED
FIRE	♈ Kali Phos	♐ Silica	♌ Mag Phos
AIR	♎ Nat Phos	♊ Kali Mur	♒ Nat Mur
WATER	♋ Calc Flor	♓ Ferr Phos	♏ Calc Sulp
EARTH	♑ Calc Phos	♍ Kali Sulp	♉ Nat Sulp

ASTROLOGY, Traditional Cell Salts

Stage 1

		Forces			Substances		
Cosmic	Nat Mur	O	♒	♄1	♄2	♑	Al	Calc Phos
	Ferr Phos	Cl	♓	♃1	♃2	♐	Mg	Silica
	Kali Phos	Si	♈	♂1	♂2	♏	C	Cal Sulp
Earthly	Nat Sulp	N	♉	♀1	♀2	♎	Ca	Nat Phos
	Kali Mur	S	♊	☿1	☿2	♍	Na	Kali Sulp
	Calc Flor	P	♋	☽1	☽2	♌	H	Mag Phos
		Being			Manifestation		

Lievegoed – Atkinson - Hauschka - Trad Cell Salts

Stage 2

Metabolic

Earthly Forces	Calc Flor	P	♋	☽1	♄2	♑	Al	Calc Phos	**Cosmic Matter**
	Kali Mur	S	♊	☿1	♃2	♐	Mg	Silica	
	Nat Sulp	N	♉	♀1	♂2	♏	C	Cal Sulp	
Cosmic Forces	Kali Phos	Si	♈	♂1	♀2	♎	Ca	Nat Phos	**Earthly Matter**
	Ferr Phos	Cl	♓	♃1	☿2	♍	Na	Kali Sulp	
	Nat Mur	O	♒	♄1	☽2	♌	H	Mag Phos	

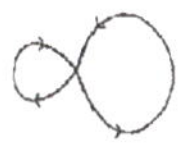

Nerve Sense

Lievegoed - Atkinson - Horizontal Spin
Trad. Cell Salts, Hauschka - RS. Ag. Course
3.50am 6.2.08 Dornach

Stage 3

Metabolic

Cosmic Matter	Calc Phos	Al	♑	♄2	☽1	♋	P	Calc Flor	**Earthly Forces**
	Silica	Mg	♐	♃2	☿1	♊	S	Kali Mur	
	Cal Sulp	C	♏	♂2	♀1	♉	N	Nat Sulp	
Cosmic Forces	Kali Phos	Si	♈	♂1	♀2	♎	Ca	Nat Phos	**Earthly Matter**
	Ferr Phos	Cl	♓	♃1	☿2	♍	Na	Kali Sulp	
	Nat Mur	O	♒	♄1	☽2	♌	H	Mag Phos	

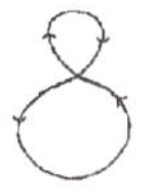

Nerve Sense

Lievegoed - Atkinson - Hauschka
Trad. Cell Salts, - Seasons

South

Whittni Grubaugh's suggestions

Nat Mur	Regulates water uptake and prevents drought stress.
Nat Phos	Neutralizes soil acidity and improves nutrient availability.
Nat Sulp	Removes excess water and toxins from soil, enhancing root health.
Kali Mur	Clears blockages in nutrient transport and helps repair damage.
Kali Phos	Supports root strength and resilience under stress.
Kali Sulp	Enhances energy transfer, improving fruit and flower production.
Calc Flour	Supports cell wall integrity, aiding in protection from diseases and pest damage.
Calc Phos	Facilitates pH balance during root and shoot development.
Calc Sulp	Detoxifies soil and assists in the healing of wounds caused by pests or pruning.
Mag Mur	Supports root-to-shoot communication, improving stress resistance.
Mag Sulp	Promotes chlorophyll production and detoxifies soil.
Mag Phos	Relieves stress in plant tissues, aiding in recovery from weather or pest damage.
Ferr Phos	Increases chlorophyll production, supporting photosynthesis and oxygenation
Silica	Enhances structural integrity, improving resistance to lodging and fungal attacks.

Sign	Salt	Food
ARIES March 21-April 20	Potassium Phosphate AKA. Kali Phos.	White beans, cucumbers, almonds, spinach, hazelnuts, lentils, avocados, kale.
TAURUS April 21-May 20	Sodium Sulphate AKA. Nat Sulph.	Lentils, spinach, peppers, paprika, pumpkins, celery, leeks, spring onions.
GEMINI May 21-June 20	Potassium Chloride AKA. Kali Mur.	Cucumber, hazelnuts, lentils, spinach, sesame seeds, potatoes, carrots, apples.
CANCER June 21-July 22	Fluoride of Lime AKA. Calc Fluor.	Raw vegetables, sesame seeds, spinach, broccoli, mushrooms, squash, pineapples.
LEO July 23-August 22	Magnesium Phosphate AKA. Mag Phos.	Brazil nuts, white beans, corn, walnuts, peas, bananas, plums, limes, gooseberries.
VIRGO August 23-Sept 22	Potassium Sulphate AKA. Kali Sulph.	Hazelnuts, almonds, spinach, lentils, peas, lettuce, flax seeds, lemons.
LIBRA Sept 23-Oct 22	Sodium Phosphate AKA. Nat Phos.	Lentils, asparagus, spinach, rose hips, olives, carrots, basil, mint, peaches.
SCORPIO Oct 23-Nov 22	Calcium Sulphate AKA. Calc Sulph.	Almonds, cucumbers, lentils, cauliflower, leeks, onions, turnips, brusselsproats.
SAGITTARIUS Nov 23-Dec 21	Silica AKA. Silicea.	Cucumbers, peas, carrots, strawberries, parsley, cabbage, nettles, apricots.
CAPRICORN Dec 22-Jan 19	Lime Phosphate AKA. Calc Phos.	Almonds, cucumbers, white beans, dandelions, cherries, spinach, dates.
AQUARIUS Jan 20-Feb 19	Sodium Chlorine AKA. Nat Mur.	Reed beets, radishes, tomatoes, celery, figs, pecan nuts, oregano, sauerkraut.
PISCES Feb 20-March 20	Iron Phosphate AKA. Ferrum Phos.	Spinach, hazelnuts, sesame seeds, tomatoes, blueberries, currants, garlic.

Practical Experiences

I have done very little physical trials of the Cell Salts on plants. I have been working with the BD preps and all the Chemical Elements, so have carried this study on as a side interest. I am wanting to use these remedies as a system, at some stage.

The one experience I have had is with **Sodium Sulphate** as a remedy for curly leaf in Peaches. I have used this over the last three spring periods with very good success.

I began spraying just before bud break, and then monthly till the attack period is over. So up to 4 applications. I use homeopathic Sodium Sulphate at 2mks per liter of water, in a motorised backpack sprayer.

I had the insight that curly leaf is an astral flowering process that has slipped into the leafing zone. Curly leaf is taking place during the flowering period. The 'curly leaf', with its deformities and coloring of the leaves, are images of this astral intrusion. So there is a loose boundary between the two spheres.

The Sodium as the Spirit's champion, who forms colloids and boundaries, directs the two players to their respective tasks. The Sulphur allows for the movement of the bodies to occur easily, as it acts as 'oil' in protein formation. With Sulphur working from the Physical, Etheric side of the game, it suggests it is the Etheric that is lacking strength and order. In past years only a copious spraying with seaweed, has had a major influence on curly leaf. Indicating nutrition - strengthening the Etheric - to be part of the answer. Here the Sodium is concentrating and ordering the Etheric activity available to the plant. Naturally with more good leaves for longer, the trees have become healthier, each season.

Brown Rot was controlled with BdMax FG4 .

An Alternative Thought

The traditional Cell Salts outlined so far has the 'accepted' relationship of the Constellations to them. While I agree with most of these relationships, **there are two I suggest can be changed.**

I suggest that **Aries be allotted with Calcium Phosphate (Calc Phos) and Capricorn be allotted with Potassium Phosphate (Kali Phos).**

The general reasons given for Aries to be associated with Kali Phos, are that Aries governs the head and thus the brain, and that Kali Phos is THE brain function cell salt and thus must be related to Aries. Capricorn, being a Saturn sign, is often associated with the skeleton, bone structure and the knees in particular, and thus Calc Phos, the bone cell salt, must be related to it.

However, it can be said that Aries is NOT particularly well renown for actually using their brains and being analytical, about anything. As a Cardinal Fire Sign ruled by Mars it is more renowned for rushing forward into action and thinking later, actually butting its skull against that of another, before engaging the brain to consider if it was a wise move or not. The skull which is a significant aspect of the head is made of Calcium Phosphate.

Capricorn on the other hand, is the Corporate executive who lives at the pinnacle of his corporate empire. He controls the structural form of the mountain he has climbed and built brick by brick. Putting together and maintaining this empire requires an immense amount of brain power to keep all aspects, in minds eye and under control. Maintaining all the many parts of their empires creates the 'brain lag' indicated for Kali Phos. This suggests Capricorns are the one who are the more likely to overwork the brain and be in need of Kali Phos.

For these reasons , but not them alone I suggest these two rulership's be reversed, thus providing the following diagram

The Patterning of the Cell Salts — an observation

There is another very good reason for making this change. In my studies, I have come to appreciate that there is a

Holographic Archetypal patterning to life and its manifestations. Order and Pattern have shown themselves to be synonymous with truth, effectiveness and an indication of the inherent relationship between the parts.

Naturally I am drawn to the Cell Salts because there are twelve of them, which immediately suggests they will organise themselves and can be thus further organised, understood and applied according to zodiacal patterns.

So the first step is to change over these two salts Kali Phos and Calc Phos. Then place them onto the zodiac circle and see what organisation immediately arises.

The first things to observe are that there is four groups of cations with three anions.

A Calcium group that has a Phosphate, Sulphate and Fluoride (a brother of Chloride)

A Potassium group that has a Phosphate, Sulphate and Chloride

A Sodium group that has a Phosphate, Sulphate and Chloride.

This leaves a fourth group which is combination of Mag Phos, Ferrum Phos and Silica. With chlorophyll production being based upon Magnesium, rather than iron as in Haemoglobin production, this could become a group of Mag Phos, Mag Sulp and Mag Cl.

On closer inspection, it becomes apparent that each part of these groups are placed in a specific relationship to the other parts of the group. Each group has the same angular relationships present within it. There is a 90 degree, 120 degree and 150 degree relationship present between each set of compounds.

Angular relationships between the planets are called aspects. Aspects between the planets are very important, as they indicate the relationship, or the quality of the 'conversation', between the planets. The planets are continually moving and therefore continually changing their degree relationship to one another, and therefore their harmonic relationship to each other continually

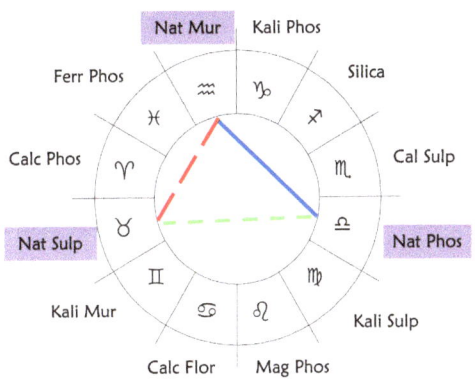

The Four Groups of Elements & Zodiacal Aspects

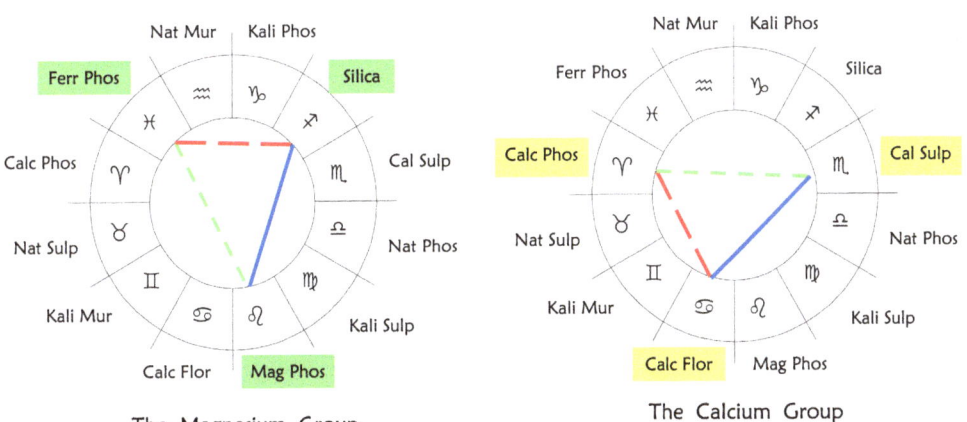

– – – Square 90 degrees — An aspect of tension and bring things into 'relief' or reality

· · · · Quinqunx 150 degrees — An aspect of movement and anticipation of change

——— Trine 120 degrees — An aspect of positive support and direction

changes. The best way to understand the nature of the aspects, is to see them as part of a complete cycle of expansion, and contraction. Not unlike Lievegoed's 'Being' and "Manifestation" processes, we see, a beginning which builds to a culmination of an impulse at the 180 degree, 'opposition' point, which is then followed by a 'coming into manifestation' phase as the contraction phase, back to zero degrees takes place. The 90 degree or square aspect is renowned as the place where an impulse reaches its first point of reality or crisis. It meets its first challenge, and is checked in its growth in someway, by an obstacle. Based upon this awareness, of this inherent fault in the process, changes are made.

At the 120 degree or trine phase, the lessons of the square have been learnt, the changes made and the process gets a very positive affirmation of the clear direction forward.

The 150 degrees stage or quinqunx, is the stage before the 180 degree opposition. The opposition is the point furtherest from the beginning, and so indicates the stage where the impulse will have expanded to its greatest point, of this cycle. This is the stage where a clear boundary is set, and the rest of the cycle will be spent bringing this defined potential of the opposition time, into being. Further growth can come in the next cycle but at the opposition, this is the boundary for this cycle. The 150 degree phase just before the 180 degree phase, is generally experienced as a time of awakening to further pending problems, after the positive affirmation of the trine. They are not so strong as to cause change, but they do cause an uneasy awareness something is coming. The opposition is the point of the change having to be addressed. So the 150 degrees is a point of uneasy movement, or movement bought about by something on the scale of an irritation.

So to summarise the 90 degree angle **sets a block** for something, the 150 degree gives a **sense of movement**, while the 120 degree angle provides a positive **direction forward.**

This threefold image is similar to that provided in Astrology by the modes Fixed elements set blocks, Mutable elements like to stay in movement while Cardinal elements like a clear direction forward.

A further observation from these diagrams is the order of the patterns. There are two sets of two orders in these patterns. One set— the Sodium (Na) and Magnesium (Mg) - has the trine leading the square, while the other— The Potassium (K) and Calcium (Ca) - has the square leading the trine. It can be further noted that Na and Mg are on the same ring (Cosmic Physical) of the Periodic Table as each other, as are Ca and K, which sit on the Cosmic Etheric ring. This mimics the polarity between the layers.

Without the changing of Calc Phos for Kali Phos to their constellation association, this patterning and order is not apparent.

How big a change is it. Both have Phos anions and they are both on the 'cosmic etheric ring'. Calcium is an element of the Internal Etheric arm, while Potassium is on the World Etheric arm. On the 'manifest' level of the PT, K is on the cation astral arm, while Ca is on the cation etheric arm.

SO they are in fact very similar, both 'directional' cardinal elements, both concerned with the Etheric's' activity yet K fosters it into manifestation through stimulating the Astral, while Ca does it directly in the etheric. K by working through the World sphere and the Astrality, talks of working in a more external and active manner than Ca. The same can be said of Aries and Capricorn. In the Seasonal Zodiac, Aries is the beginning and the individual alone, while Capricorn is the 10th sign of culmination and worldly success . In 'Lievegoed zodiac' Aries is Mars 1, a stage where the cosmic impulse comes into the etheric sphere, while Capricorn is Saturn 2 or the stage where this Mars 1 activity finally reaches seed development.

Atkinson altered Trad. Cell Salts
Calc Phos and Kali Phos changed

	CARDINAL	MUTABLE	FIXED
FIRE	♈︎ Calc Phos	♐︎ Silica	♌︎ Mag Phos
AIR	♎︎ Nat Phos	♊︎ Kali Mur	♒︎ Nat Mur
WATER	♋︎ Calc Flor	♓︎ Ferr Phos	♏︎ Calc Sulp
EARTH	♑︎ Kali Phos	♍︎ Kali Sulp	♉︎ Nat Sulp

ASTROLOGY, Altered Traditional Cell Salts

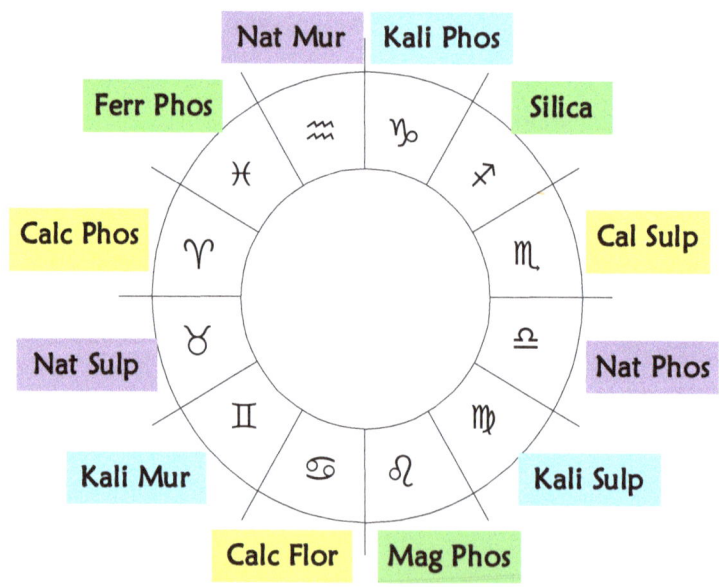

Traditional Cell Salts ala Atkinson

Stage 1

		Forces			Substances		
Cosmic	Nat Mur	O	♒	♄1	♄2 ♑	Al	Kali Phos
	Ferr Phos	Cl	♓	♃1	♃2 ♐	Mg	Silica
	Calc Phos	Si	♈	♂1	♂2 ♏	C	Cal Sulp
Earthly	Nat Sulp	N	♉	♀1	♀2 ♎	Ca	Nat Phos
	Kali Mur	S	♊	☿1	☿2 ♍	Na	Kali Sulp
	Calc Flor	P	♋	☽1	☽2 ♌	H	Mag Phos
		Being			**Manifestation**		

Lievegoed – Atkinson – Hauschka 1 – Atk. Trad Cell Salts

Stage 2

Metabolic

Earthly Forces	Calc Flor	P	♋	☽1	♄2 ♑	Al	Kali Phos	**Cosmic Matter**
	Kali Mur	S	♊	☿1	♃2 ♐	Mg	Silica	
	Nat Sulp	N	♉	♀1	♂2 ♏	C	Cal Sulp	
Cosmic Forces	Calc Phos	Si	♈	♂1	♀2 ♎	Ca	Nat Phos	**Earthly Matter**
	Ferr Phos	Cl	♓	♃1	☿2 ♍	Na	Kali Sulp	
	Nat Mur	O	♒	♄1	☽2 ♌	H	Mag Phos	

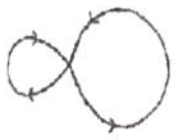

Nerve Sense

Lievegoed - Atkinson - Horizontal Spin
Atk. Trad. Cell Salts, Hauschka 5 - RS. Ag. Course
3.50am 6.2.05 Dornach

Stage 3

Metabolic

Cosmic Matter	Kali Phos	Al	♑	♄2	☽1 ♋	P	Calc Flor	**Earthly Forces**
	Silica	Mg	♐	♃2	☿1 ♊	S	Kali Mur	
	Cal Sulp	C	♏	♂2	♀1 ♉	N	Nat Sulp	
Cosmic Forces	Calc Phos	Si	♈	♂1	♀2 ♎	Ca	Nat Phos	**Earthly Matter**
	Ferr Phos	Cl	♓	♃1	☿2 ♍	Na	Kali Sulp	
	Nat Mur	O	♒	♄1	☽2 ♌	H	Mag Phos	

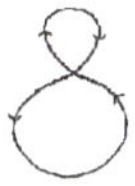

Nerve Sense

Lievegoed - Atkinson - Hauschka
Atk. Trad Cell Salts, - Seasons

South

Atkinson 2 Cations and Anions

The question of **Which Cation is allotted to which Energetic Activity**, is still under discussion. I have collected 7 alternative suggestions. Each of these can be moved through the Three Worlds, as done on the previous page. What is your suggestion?

		CARDINAL	MUTABLE	FIXED
		Cl	S	P
FIRE	Na	♈ Nat Mur	♐ Nat Sulp	♌ Nat Phos
AIR	K	♎ Kali Mur	♊ Kali Sulp	♒ Kali Phos
WATER	Ca	♋ Calc Flour	♓ Calc Sulp	♏ Calc Phos
EARTH	Mg	♑ Mag Mur Silica	♍ Mag Sulp Fe Pos	♉ Mag Phos

ASTROLOGY, Elements & Modes, **Atkinson Cations & Anions,** Cell Salts
Na, K , Ca, Mg

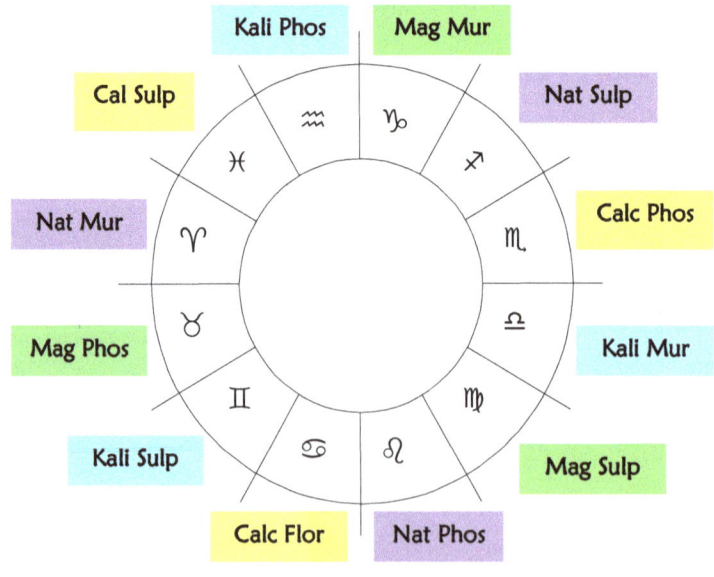

Hauschka Cations
Pg 73 'Nutrition'

		CARDINAL	MUTABLE	FIXED
		Cl	S	P
FIRE	Mg	♈	♐	♌
		Mag Mur	Mag Sulp	Mag Phos
AIR	Na	♎	♊	♒
		Nat Mur	Nat Sulp	Nat Phos
WATER	K	♋	♓	♏
		Kali Mur	Kali Sulp	Kali Phos
EARTH	Ca	♑	♍	♉
		Calc Mur	Calc Sulp	Calc Phos

ASTROLOGY, Elements & Modes, Cations & Anions, Cell Salts
Na, K, Ca, Mg

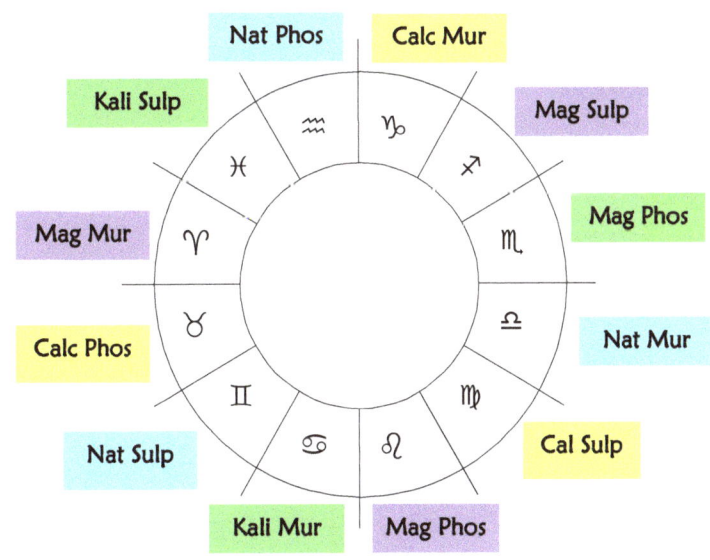

Zodiacal References
Hauschka Cations
Astro Anions

Hausmann - Klett Cations

		CARDINAL	MUTABLE	FIXED
		Cl	S	P
FIRE	Na	♈	♐	♌
		Nat Mur	Nat Sulp	Nat Phos
AIR	Mg	♎	♊	♒
		Mag Mur	Mag Sulp	Mag Phos
WATER	K	♋	♓	♏
		Kali Mur	Kali Sulp	Kali Phos
EARTH	Ca	♑	♍	♉
		Cal Mur	Cal Sulp	Cal Phos

Elements & Modes, Cations & Anions, Cell Salts
Hausmann Klett Na, Mg , K , Ca

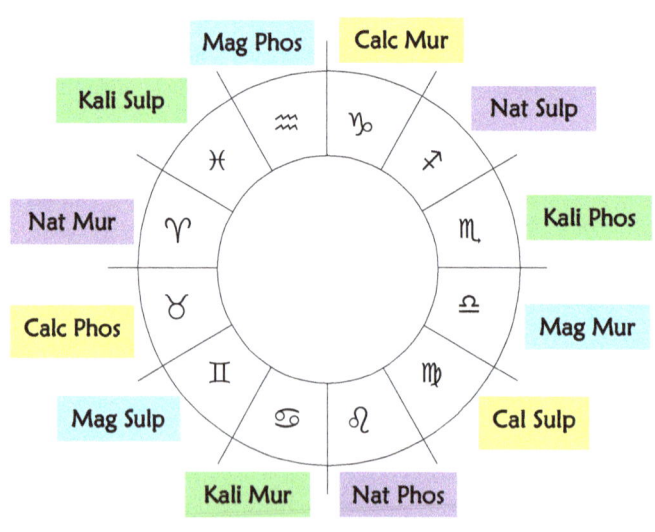

Zodiacal References
Hausmann Cations
Astro Anions

Bacchus - Steens Cations

		CARDINAL	MUTABLE	FIXED
		Cl	S	P
FIRE	Mg	♈︎ Mag Mur	♐︎ Mag Sulp	♌︎ Mag Phos
AIR	K	♎︎ Kali Mur	♊︎ Kali Sulp	♒︎ Kali Phos
WATER	Na	♋︎ Nat Mur	♓︎ Nat Sulp	♏︎ Nat Phos
EARTH	Ca	♑︎ Cal Mur	♍︎ Cal Sulp	♉︎ Cal Phos

Elements & Modes, Cations & Anions, Cell Salts
Mg , K , Na, Ca

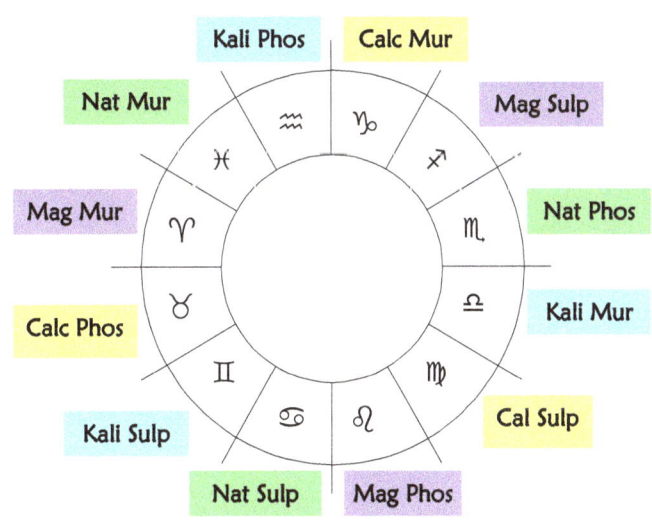

Zodiacal References
Atkinson 3 Cations
Astro Anions

R Splawn Cations & Anions

		CARDINAL	MUTABLE	FIXED
		P	S	Cl
FIRE	K	♈	♐	♌
		Kali Phos	Kali Sulp	Kali Mur
AIR	Mg	♎	♊	♒
		Mag Phos	Mag Sulp	Mag Mur
WATER	Na	♋	♓	♏
		Nat Phos	Nat Sulp	Nat Mur
EARTH	Ca	♑	♍	♉
		Cal Phos	Cal Sulp	Cal Mur

Elements & Modes, Cations & Anions, Cell Salts
K , Mg, Na, Ca

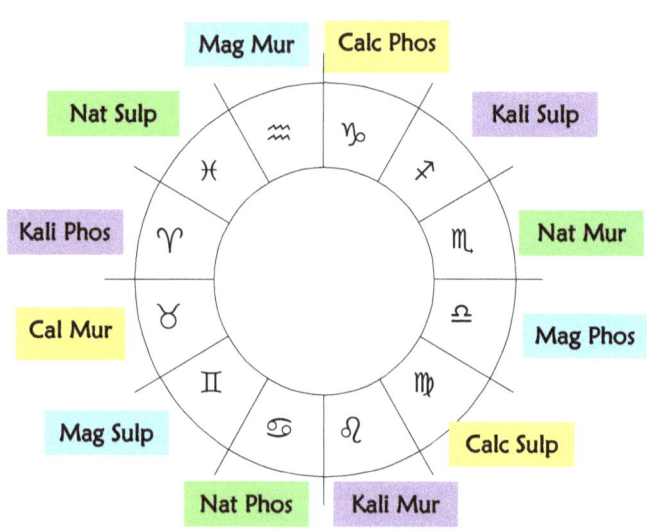

Zodiacal References
Splawn Cations
Splawn Anions

Glen One Cations
Na, K, Mg, Ca

		CARDINAL	MUTABLE	FIXED
		Atmosphere	Hydrosphere	Geosphere
		Cl	S	P
FIRE		♈	♐	♌
	Na	Nat Mur	Nat Sulp	Nat Phos
AIR		♎	♊	♒
	K	Kali Mur	Kali Sulp	Kali Phos
WATER		♋	♓	♏
	Mg	Mag Mur	Mg Sulp	Mag Phos
EARTH		♑	♍	♉
	Ca	Calc Mur	Calc Sulp	Calc Phos

ASTROLOGY
Spheres to Modes

Atkinson One

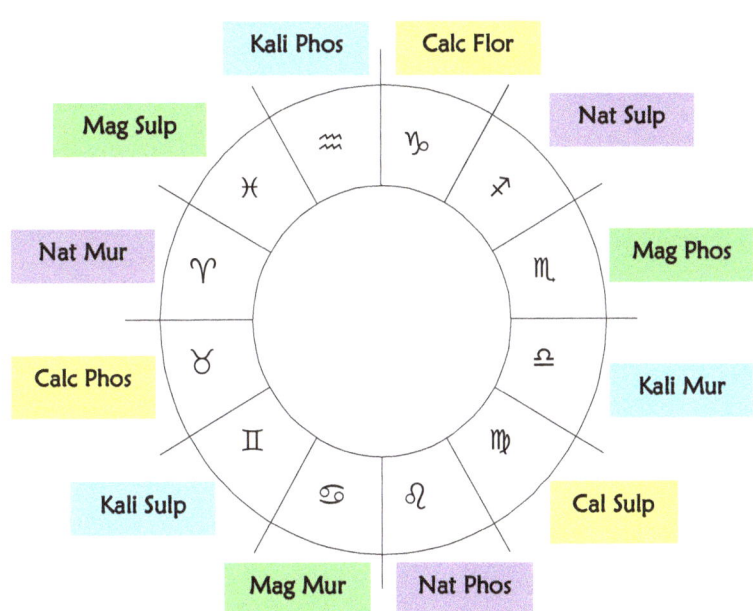

Moodie Cations and Anions
Cations Mg, Na, Ca, K
Anions P, S, Cl

		CARDINAL Atmosphere P	MUTABLE Hydrosphere S	FIXED Geosphere Cl
FIRE		♈︎	♐︎	♌︎
	Mg	Mag Phos	Mag Sulp	Mag Mur
AIR		♎︎	♊︎	♒︎
	Na	Nat Phos	Nat Sulp	Nat Mur
WATER		♋︎	♓︎	♏︎
	Ca	Calc Phos	Calc Sulp	Calc Flor
EARTH		♑︎	♍︎	♉︎
	K	Kali phos	Kali Sulp	Kali Mur

ASTROLOGY
Spheres to Modes
Cell Salts

Moodie Cations
Moodie Anions
to Sphere & Mode

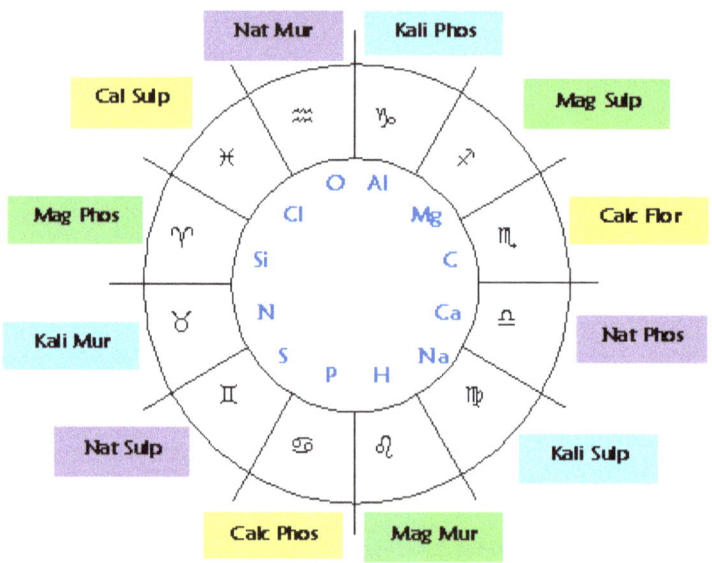

Zodiacal References

Moodie Cations
Moodie Anions

Hauschka
original

180

Hauschka and Julius

As part of working with the chemical elements in agriculture, one is drawn to the question of what the relationship of the chemical elements is to the 12 fold structure of organisation. Both Dr Steiner (RS) and Dr Hauschka (RH) gave firm indications that these elements are related to and indeed are 'sourced' from the constellational star activity and thus the Galactic sphere. These are star forces that have been accumulated as energetic standing waves in the Earth.

Dr Hauschka made very specific indications in his "Nature of Substance" of these relationships, while Dr Steiner suggested the mineral kingdom was the best tool for effecting the Spirit's function of Humans and Animals. The Galactic Sphere is the natural home / source of the Spirit and its 'architectural' fundamental plan activity. There are 12 constellations and so how can chemistry fall into this archetype.

One problem with this question is that there are 122 chemical elements on the Periodic Table and only 12 constellations. Hauschka does an admirable job of providing a picture of the relationships between the major chemical elements, however his organisation does not mesh easily with the structure of the traditional Periodic Table, and does not cover all of the elements. For example he allocates separate constellations for Ca and Mg, and N and P, which are in the same chemical elemental groups respectively, however he lumps together all Alkalis, K and Na as one group, and all Chlorides, F,Cl, I as another. So he is providing a partial view of the subject focusing on some primary elements of Life and real world observations of their place in existence. All very valuable, however it does not encompass all of chemistry. I have offered this on my 'Glenological Chemistry', which approaches this subject from a different observation.

The highlight of Hauschka's book is the archetypal order, he shows in his division of his 12 primary elements into the three 'spheres' of

	CARDINAL	MUTABLE	FIXED
Aristotle	Atmosphere	Hydrosphere	Geosphere
FIRE	♌	♊	♋
	H	S	P
AIR	♉	♓	♈
	N	Cl	Si
WATER	♒	♐	♑
	O	Mg	Al
EARTH	♏	♍	♎
	C	Na	Ca

HAUSCHKA 1

ASTROLOGY
Spheres to Modes

Hauschka Pg 156
Constellations

Aristotle

Geosphere, Hydrosphere and Atmosphere, and the three 'crosses' of the Cardinal, Mutable and Fixed constellations. This is a great observation and the one thing from RH I keep all the way through.

Chemicals in the Earthly Spheres is Good

Things can become a little confusing though. He has two references

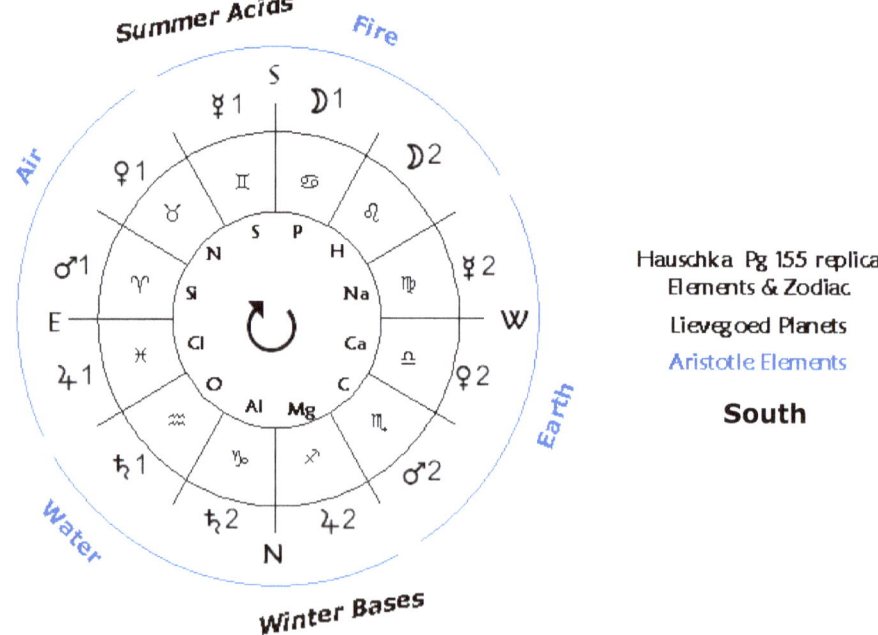

Hauschka Pg 155 replica
Elements & Zodiac.

Lievegoed Planets

Aristotle Elements

South

182

running. The Earthly Spheres he uses to establish his Atmosphere Hydrosphere and Geosphere crosses, mimics the Astrological Modes. However he has the traditionally

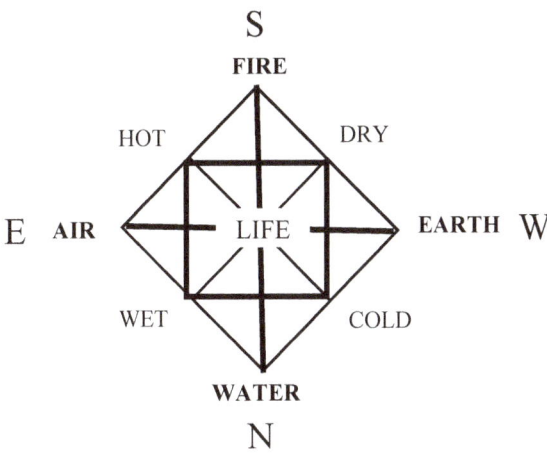

Cardinal Constellations aligned with the obviously Fixed Geosphere, and the Fixed Constellations with the Atmosphere, which is a Cardinal reference. He then uses Aristotle's Elements, with a grouping of one of each of the members of the Earthly Spheres.

Hence his elemental reference of the constellations does not

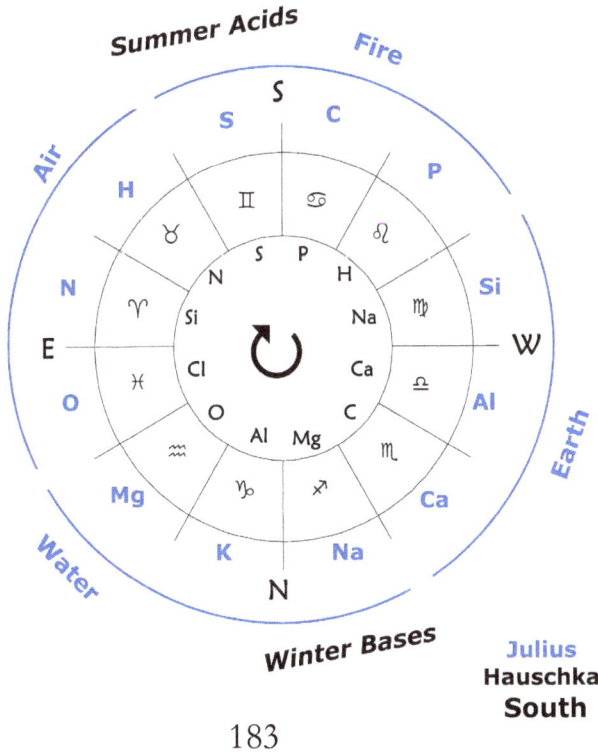

conform to the earlier Persian allocations that are commonly used today. An Aristotelian elemental zone encompasses three adjoining constellations. Rather than each constellation having its own individual element.

His allocation of the modes to his three spheres do not conform to what would be considered the obvious relationships, based on the constellations. Also the natural order of the constellations is disturb between his picture on p155 and the diagram on pg. 156. These peculiarities however provide opportunities for further investigations.

I have not yet found a direct reference of RH's order to the Circular 122 element Periodic Table, I have developed. This does not negate either view. It indicates two different 'wormholes' of exploration of the same subject. RH is looking at the primary 12 fold archetype behind all the elements, while I am talking of all of chemistry as one whole. These are two different dimensions of organisation, so the rules and associations can be very different, in each view.

Another RS influenced view of Chemistry is presented by Fritz Julius in his book, "Fundamentals for a Phenomenological Study of Chemistry". Julius is a chemistry school teacher. He has completely different reasons for his associations, which make very good sense. He appears to have made a more direct reference to Aristotle's Elements, and the Chemical Elements he puts in each group bear more commonality to the Aristotelian Element than Hauschka's do.

Again, not making either wrong here, just noting their differences

To find the relationship between the two we can use the Aristotle elements both use, and **add the constellations to Julius.**

In the discussion of RS view of the chemical elements there is also a reference available through Husemann in his second volume of "Anthroposophical Approach to Medicine".

Hauschka

While reading Nature of Substance I was continually dismayed by RH's use of seasonal references for determining correspondences to the zodiac. Being from the southern hemisphere, were the seasons are the opposite to those of the northern hemisphere, I am drawn to ask, What about us? How do these elements relate to the seasons as they occur in the southern hemisphere?

Hauschka makes reference to ancient wisdom and uses Aristote's reference to the elements, used commonly by 'alchemists'. His arguments are very convincing and there is a lot of finesse in this model. No doubt his associations are correct and have value within certain spheres.

I would however like to present another set of zodiacal references for his elements. These are presented not to make Hauschka wrong, but to provide a different view on the same information. It is as if Hauschka has made one conclusion that leads us up a certain road, yet if we use some of his conclusions, and take a different fork in the road, we will be lead down a different path, providing a different conclusion that can provide us a window into another dimensional reality, and other possible applications.

Aristotle	CARDINAL Atmosphere	MUTABLE Hydrosphere	FIXED Geosphere
FIRE	♌ H	♊ S	♋ P
AIR	♉ N	♓ Cl	♈ Si
WATER	♒ O	♐ Mg	♑ Al
EARTH	♏ C	♍ Na	♎ Ca

HAUSCHKA 1

ASTROLOGY
Spheres to Modes

Hauschka Pg 156
Constellations

Aristotle

The outcomes are not mutually exclusion, rather they are as if they are siblings, each being right in their own way, but having different applications in life.

Towards the end of his book RH has followed particular processes. He has outlined three groups of four elements and related each set of elements to the organic sphere in which they predominate. H,N,O,C are to be found in the Atmosphere, Na, Cl, Mg, S are found predominately within the oceans or Hydrosphere, while Ca, Si, Al and P are found predominately in the Earth, or Geosphere.

Each of these four elements in each group has been identified as carrying a particular relationship to one of the four elements Fire, Air, Water or Earth. These elemental allocations however are from Aristotle and are different to the Persian ideals of the elements and the constellations. This offers an opportunity.

	CARDINAL Atmosphere	**FIXED** Geosphere	**MUTABLE** Hydrosphere	
FIRE	♌ / H	♋ / P	♊ / S	
AIR	♉ / N	♈ / Si	♓ / Cl	**Aristotle** / **Hauschka 2**
WATER	♒ / O	♑ / Al	♐ / Mg	**Persian Modes**
EARTH	♏ / C	♎ / Ca	♍ / Na	
	Fixed	Cardinal	Mutable	

On page 156 of 'Nature of Substance' RH presents this diagram showing all these relationships. A student of Astrology will note that this patterning is exactly the same as the way in which the Zodiacal references are made with regards to the threefold modes and the

		CARDINAL	MUTABLE	FIXED
		Atmosphere	Hydrosphere	Geosphere
		Cl	S	P
FIRE		♋	♊	♌
	Mg	H	S	P
AIR		♈	♓	♉
	Na	N	Cl	Si
WATER		♑	♐	♒
	K	O	Mg	Al
EARTH		♎	♍	♏
	Ca	C	Na	Ca

HAUSCHKA 3

Aristotle Elements to Constellations

Astrology Modes to Spheres to Constellations

Hauschka Cations
Atkinson Anions

fourfold elements, however Hauschka does not use this pattern to make a zodiacal reference from. He has generated his reference from the Aristotle Elements and his observations of the Chemicals and the Seasons.

The first thing that strikes me when looking at this chart is that **the constellational order is upset**. The Astro. Elements are in the right relationship to each other, however the **Constellations are out of order** eg in the sky we would see Gemini, Cancer, Leo, however RH has them in his chart as Cancer, Gemini, Leo. This is different to the circular diagram Hauschka has in his pg. 155

		CARDINAL	MUTABLE	FIXED
		Atmosphere	Hydrosphere	Geosphere
		Cl	S	P
FIRE		♈	♐	♌
	Mg	H	S	P
AIR		♎	♊	♒
	Na	N	Cl	Si
WATER		♋	♓	♏
	K	O	Mg	Al
EARTH		♑	♍	♉
	Ca	C	Na	Ca

HAUSCHKA 4

ASTROLOGY
Spheres to Modes

Astrology Constellations to Chemicals to Aristotle Elements

Hauschka Cations
Steiner Anions

		CARDINAL	MUTABLE	FIXED	
		Atmosphere	Hydrosphere	Geosphere	
		Cl	S	P	
FIRE		♈	♐	♌	**HAUSCHKA 5**
	Mg	Si	Mg	H	**ASTROLOGY** Spheres to Modes
AIR		♎	♊	♒	Astrology Constellations
	Na	Ca	S	O	Hauschka / 3 to Natural polarities
WATER		♋	♓	♏	Hauschka Cations Steiner Anions
	K	P	Cl	C	
EARTH		♑	♍	♉	
	Ca	Al	Na	N	

diagram with the Aristotle Elements. So what happens if we **put the constellations in their sequential order.**

It can also be noted that in this diagram when the **Astrological modes and the 3 fold RS three physical systems used by R Steiner are aligned,** and that the **Fixed mode is associated with the Persian Cardinal constellations**, and likewise for the **Cardinal mode and Fixed constellations**. This inconsistency does not make RH wrong it just suggests we have an opportunity to look for other relationships than his.

It would therefore also seem reasonable to use this age old reference system **defining the constellations by the Element and Mode of each constellation** to also **establish a zodiacal**

		CARDINAL	MUTABLE	FIXED	
		Atmosphere	Hydrosphere	Geosphere	
		Cl	S	P	
FIRE		♈	♐	♌	**HAUSCHKA 6**
	Mg	H	Mg	Si	**ASTROLOGY** Spheres to Modes
AIR		♎	♊	♒	Hauschka Spheres
	Na	O	S	Ca	Astrology Constellations To Hauschka Chemicals
WATER		♋	♓	♏	Hauschka Cations Steiner Anions
	K	C	Cl	P	
EARTH		♑	♍	♉	
	Ca	N	Na	Al	

pattern. We should keep RH's insights into the chemical elements relationship to the 4 fold elements, and to the physical organisations.

In line with RS image of the 'Agricultural individuality', I have associated the Fixed activity with the Earth, and the Atmospheric activity with the Cardinal / Metabolic processes. Thus the Mutable is related to the Oceanic Cross. This is still using the Aristotelean Elements.

Next we can apply the traditional zodiacal associations and we are provided with these relationships. (below) **This provides different constellation to Chemical Element relationships.** Note RH's chemicals are still in their original places.

A further consideration for change can be made if we **move the Chemicals to RH's original constellational ruler,** while keeping the Astrological adjustments we have made along the way. We keep the Chemicals in their Sphere groupings, and keep them with their Hauschka Constellation, but use the Astrological order. This action misaligns the Constellational Elements with the Spheres again. So what happens when these are realigned.

If we go back to Hauschka 4, where RH's Chemicals are in their original order, but within an Astrological context, we can note various reasons for **moving the Chemicals around within the**

		CARDINAL	MUTABLE	FIXED	
		Atmosphere	Hydrosphere	Geosphere	
		Cl	S	P	
FIRE		♈	♐	♌	**HAUSCHKA 7**
	Mg	H	Cl	P	ASTROLOGY Spheres to Modes
AIR		♎	♊	♒	Astrology Constellations
	Na	N	S	Si	to Chemicals to Astro Elements
WATER		♋	♓	♏	Hauschka Cations
	K	O	Mg	Ca	Steiner Anions
EARTH		♑	♍	♉	Haushka Polarities
	Ca	C	Na	Al	

groupings. In the Atmosphere group the 4 elements we identify as the 'elements of Protein' and the carriers into the physical of the spiritual activities are arranged as we would expect, in relationship to the Astro elements. This provides a **natural structure of polarity**. Carbon with Hydrogen, Oxygen with Nitrogen. However the same can not be said of the other groupings. RH talks of polarities between the elements but due to his constellational reference based upon seasonal considerations, he also places them on opposite constellations, however in pg. 156 diagram this polarity arrangement does not occur.

With regard to Hydrosphere it can be suggested that as Cl is more acidic than S, and that it sits on the World Spirit arm of the PT, while S sits on the internal Astral arm, that these two should change places. This restores the natural relationships between NaCl and MgS. Two basic stable and very useful elements in nature.

In the Geosphere group Ca and Al can be changed. Ca is the most etheric of elements in the PT, being the etheric element on all three layers of the PT, Al is the element that is on the World Physical arm on the physical ring, thus it is the more earthly element of the two. This the AlP polarity is restored as is the natural Ca/Si polarity.

With these changes there is **order between Astrology' modes and elements, Hauschka's spheres, the Periodic Table, the Agriculture Course and the Cell Salts.** This is represented on this circular diagram.

There is an alternative to consider which is an alteration to the **Geosphere Chemicals to the Energetic Elements.** What if Earth is Ca and Si is Fire, while al is water and P is Air. This is only a change of P and Si to Hauschka 1.

These patterns for the elements can be cross referenced to the Lievegoed organisations as well. While this exercise has produced many 'possibilities', it is your choice as to which you might like or find relevant.

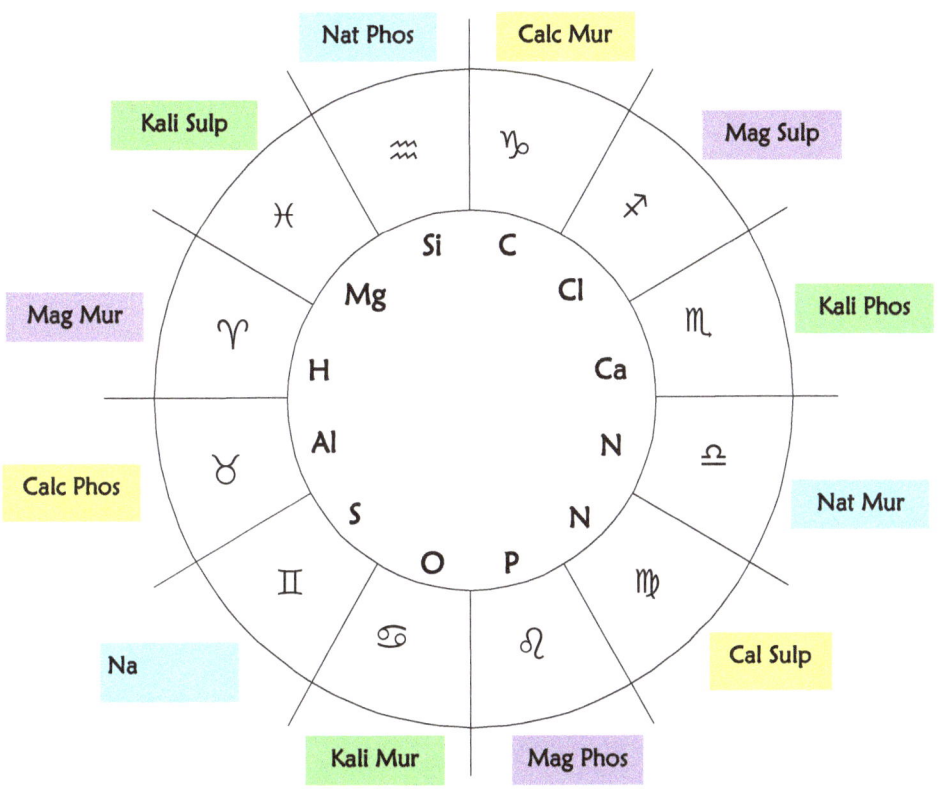

Hauschka 7
Hauschka Cell Salts
South

Following a Hauschka Thought

The two top diagrams are the beginning and an end of a story. The pic on the top left is how Dr Lievegoed (BL) outlined the 12 fold planetary order. His double planets pattern is how the planets sit in the Zodiac. This is a well accepted image, along with the Incarnating and Excarnating processes he describes, as the process of moving from Forces to Substance. To this I have added Constellations and also Dr Hauschka's chemical elements. BL's pic though is orientated to the Ecliptic of the Northern hemisphere, so it is facing South. Once we work with chemistry we must orientate off Magnetic North as this is how chemistry works. So there needs to be a horizontal flip, so the Saturn 1 planets etc. are on the right hand side of the picture. (see pg. 115)

The pic on the top right is what happens to the planets if we place the chemical elements according to Hauschka's Earth Spheres.

Forces			Substances		
O	♒	♄ 1	♄ 2	♑	Al
Cl	♓	♃ 1	♃ 2	♐	Mg
Si	♈	♂ 1	♂ 2	♏	C
N	♉	♀ 1	♀ 2	♎	Ca
S	♊	☿ 1	☿ 2	♍	Na
P	♋	☽ 1	☽ 2	♌	H

Planets — Dr. B. Lievegoed
Constellations — Mr. G. Atkinson
Chemical Elements — Dr. R. Hauschka

South Facing   **Glen R Atkinson**

A1 P	♄ 2	☽ 1	
Mg S	♃ 2	♀ 1	Metabolic
C N	♂ 2	♀ 1	
			Rhythmic
H O	☽ 2	♄ 1	
Na Cl	☿ 2	♃ 1	Nerve Sense
Ca Si	♀ 2	♂ 1	

Dr. R. Hauschka Dr. B. Lievegoed
Dr. R. Steiner Agric. Course

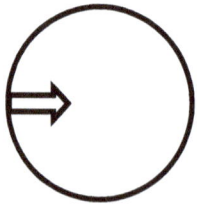

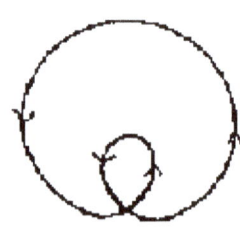

This ordering of the planets is essentially the pattern RS used within the Agriculture Course lectures. So Soul World / Stage 2.

There has always been a question for me, about how the below ground planets at Stage 2, appear after the Stage 3 flip. I have presented observations of the Seasons, of this flip as a further twist of the bottom half of the diagram. See my Stage 3 diagrams. However in 'the BIG story' the flip can also be back upon itself. If this is the case then we would find the order given in the image below.

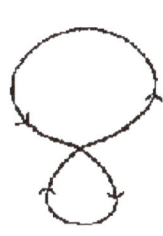

This leads to the question 'how is this ordering applicable? Is this the pattern that nature will follow through the seasonal round?' So from middle Winter we would have Mars 1, Jupiter 1, Saturn 1 Spring Equinox, Venus 1, Mercury 1, Moon 1, Summer Solstice and so on round to Venus 2. Top Right image is the simple flip bottom to the right.

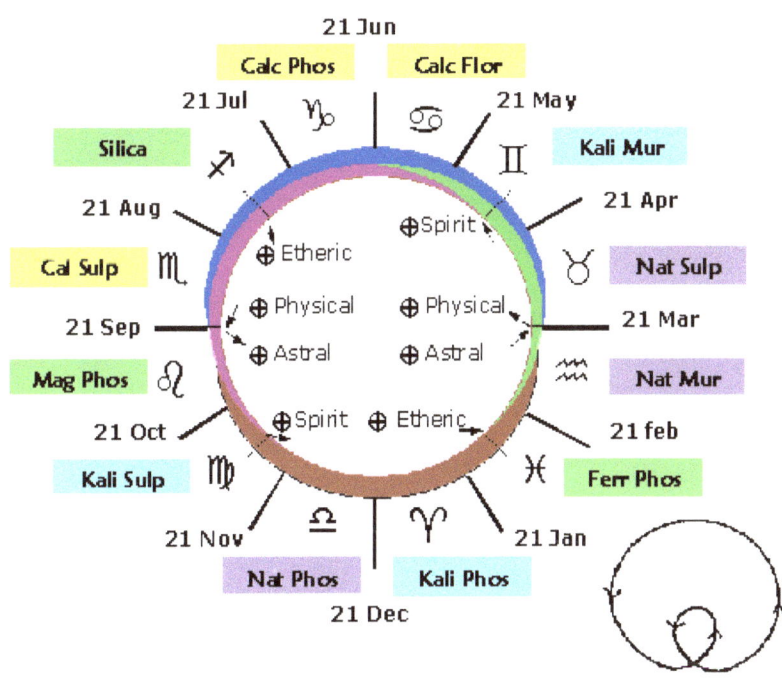

Earth Bodies - Physical World Signs - flip
Seasons - Cell Salts - Northern Hemisphere

193

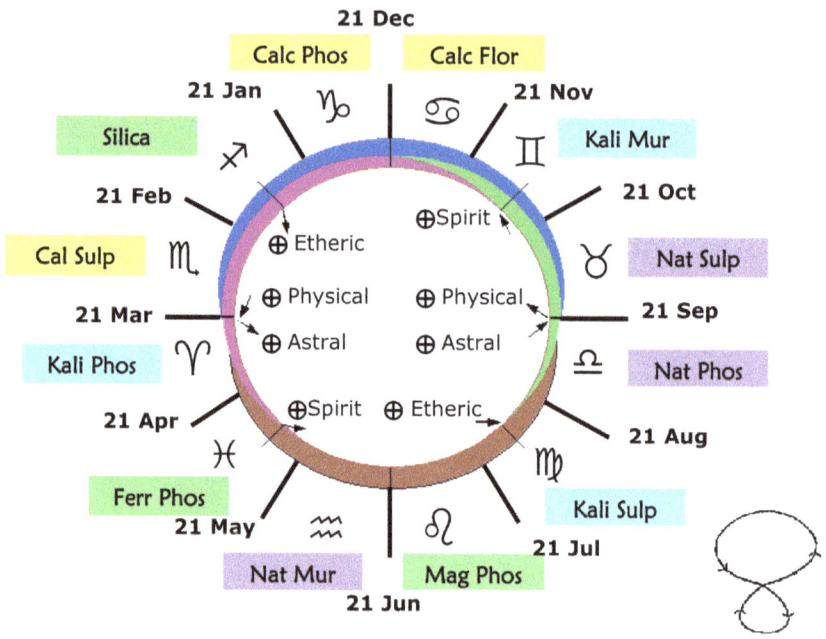

Earth Bodies - **Physical World Signs**
Seasons - Cell Salts - Southern Hemisphere

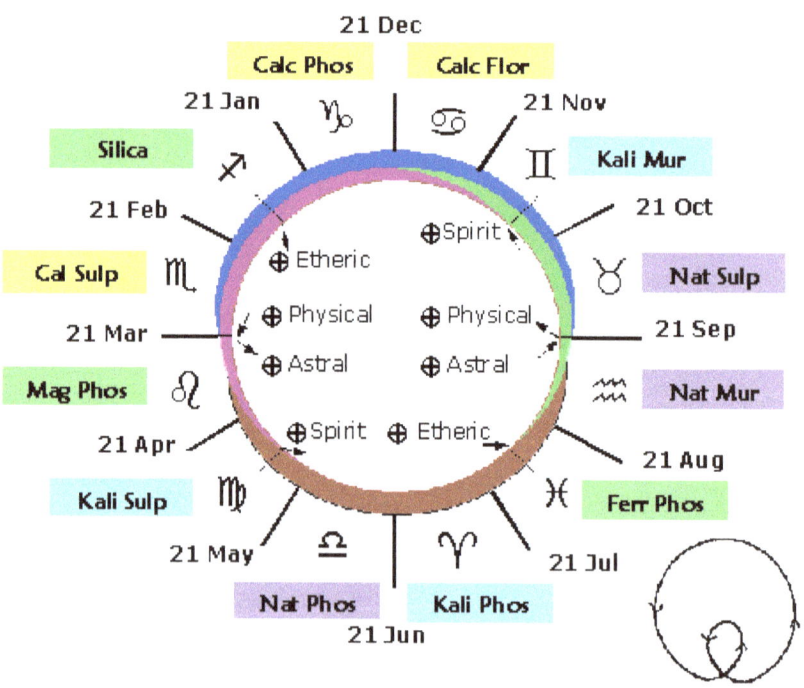

Earth Bodies - **Physical World Signs** - flip
Seasons - Cell Salts - Southern Hemisphere

Julius

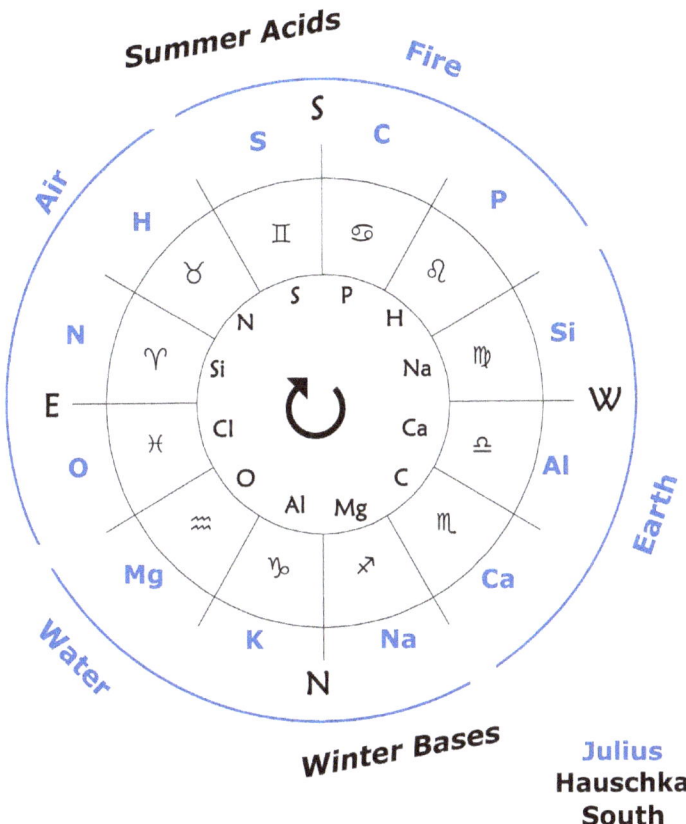

Julius
Hauschka
South

	CARDINAL	FIXED	MUTABLE
	Atmosphere	Geosphere	Hydrosphere
FIRE	♌	♋	♊
	P	C	S
AIR	♉	♈	♓
	H	N	O
WATER	♒	♑	♐
	Mg	K	Na
EARTH	♏	♎	♍
	Ca	Al	Si

JULIUS 2

Constellations
natural order

Aristotle

| | CARDINAL | MUTABLE | FIXED | |
Aristotle	Atmosphere	Hydrosphere	Geosphere	
FIRE	♌	♊	♋	**JULIUS 1**
	P	S	C	**ASTROLOGY** Spheres to Modes
AIR	♉	♓	♈	Hauschka Pg 156
	H	O	N	Constellations
WATER	♒	♐	♑	Aristotle
	Mg	Na	K	No Cl
EARTH	♏	♍	♎	
	Ca	Si	Al	

Julius 1 Elements ordered according to Hauschka's Pg 156 chart.

Julius 2 is his Chemical Elements and Aristotle's Element groups.

Julius 3 Astro Elements and Modes, Earth Spheres
Switch of Julius 2s Fixed Constellations from Atmosphere to the Geosphere
And the Cardinal Constellations from Geosphere to the Atmosphere
Using the Elements of modern Astrology for the Constellations.
With Hauschka's Cations

| | | CARDINAL | MUTABLE | FIXED | |
| | | Atmosphere | Ocean | Earth | |
		Cl	S	P	
FIRE		♈	♐	♌	
	Mg	N	Na	P	
AIR		♎	♊	♒	Hauschka Cell Salts
	Na	Al	S	Mg	**ASTROLOGY** constellations
WATER		♋	♓	♏	**JULIUS 3**
	K	C	O	Ca	
EARTH		♑	♍	♉	
	Ca	K	Si	H	

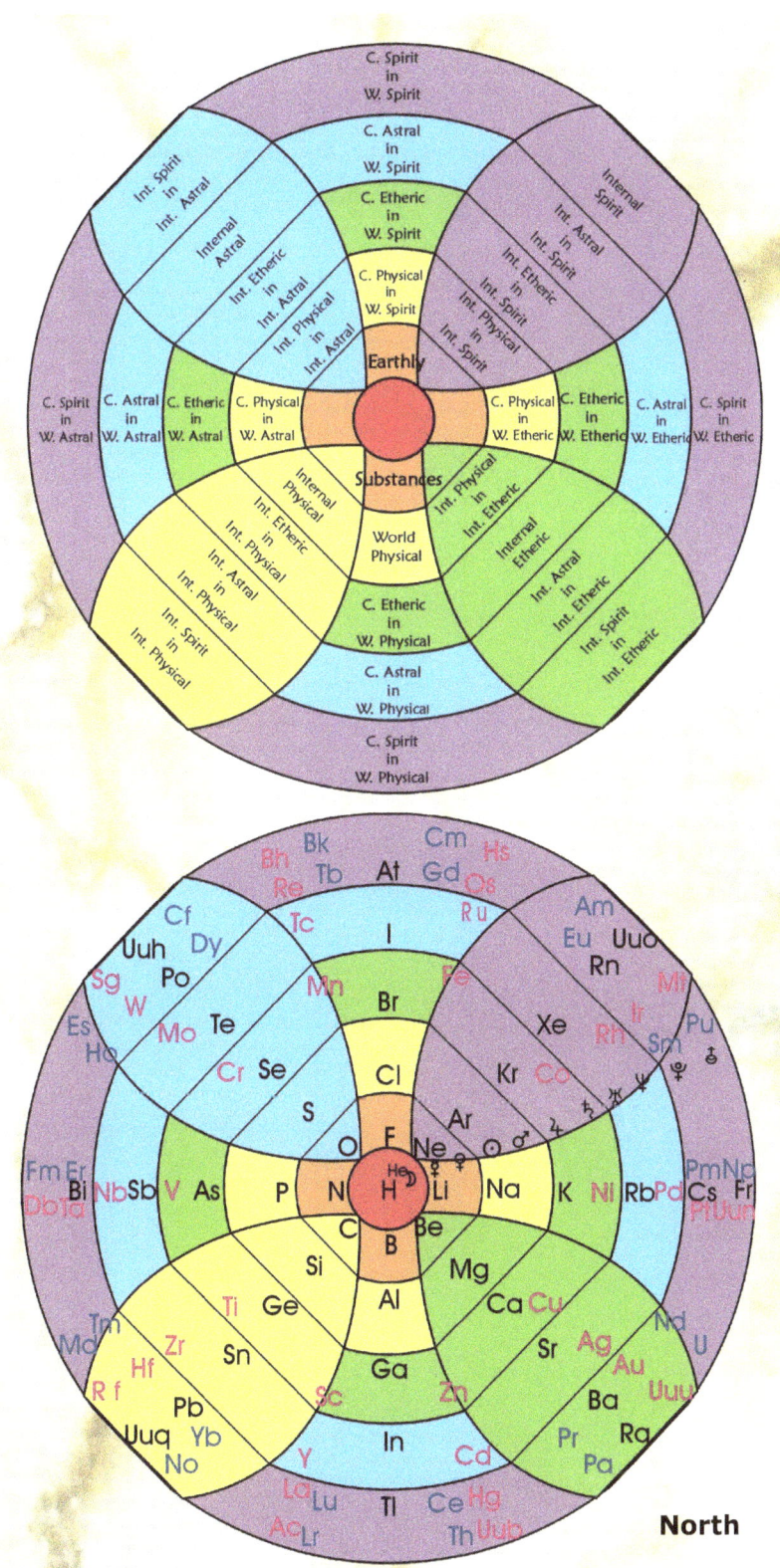

Conclusion

So much more can be said about the thought projections of the Cell Salts and Hauschka. I have not worked practically into all this enough to present anything further at this time. Its food for thought and ideas of how given information can be moved forward using existing references.

This book as a whole is a collection of questions I felt needed clarification. For myself firstly, however if they are of interest to others then all the better.

www.ingramcontent.com/pod-product-compliance
Lightning Source LLC
Chambersburg PA
CBHW041312240426
43661CB00065B/2904